BIG HISTORY

BIG HISTORY

From the Big Bang to the Present

CYNTHIA STOKES BROWN

© 2007 by Cynthia Stokes Brown
Preface to the 2012 edition © 2012 by Cynthia Stokes Brown
All rights reserved.
No part of this book may be reproduced, in any form,
without written permission from the publisher.

Every effort has been made to contact rights holders of copyrighted material. If by chance
we've neglected to provide the proper credit information, please let us know and we will
correct the omission in future printings.

Requests for permission to reproduce selections from this book should be mailed to:
Permissions Department, The New Press, 120 Wall Street, 31st floor, New York, NY 10005.

First published in the United States by The New Press, New York, 2007
This paperback edition published by The New Press, 2012
Distributed by Two Rivers Distribution

ISBN 978-1-59558-848-7 (pbk.)
ISBN 978-1-59558-845-6 (e-book)

LIBRARY OF CONGRESS CATALOGING-IN-PUBLICATION DATA
Brown, Cynthia Stokes.
Big history : from the Big Bang to the present / Cynthia Stokes Brown.
p. cm.
ISBN 978-1-59558-196-9 (hc.)
1. World history. 2. Human ecology. I. Title.
D20.B77 2007
909—dc22   2007006741

The New Press publishes books that promote and enrich public discussion and understanding of the issues vital to our democracy and to a more equitable world. These books
are made possible by the enthusiasm of our readers; the support of a committed group of
donors, large and small; the collaboration of our many partners in the independent media
and the not-for-profit sector; booksellers, who often hand-sell New Press books;
librarians; and above all by our authors.

www.thenewpress.com

Unless indicated otherwise, illustrations were created by Rob Carmichael
Composition by dix!
This book was set in Kepler

Printed in the United States of America

6 8 10 9 7 5

Recognizing that the international community is on the horns of a dilemma [the environmental impact of spectacular economic growth since 1945], world historians must look beyond development for an organizing principle. The new narrative of world history must have ecological process as its major theme. It must keep human events within the context of where they really happen, and that is the ecosystem of the earth. The story of world history, if it is to be balanced and accurate, will inevitably consider the natural environment and the myriad ways in which it has both affected and been affected by human activities.

—J. Donald Hughes, ed. *The Face of the Earth: Environment and World History*

Contents

List of Illustrations

Preface to the 2012 Edition

A lot has happened in big history in the five years since the first edition of this book appeared, in September 2007. The idea has begun to take hold. Its power of clarification is appealing to increasing numbers of people around the world. As one small example, this book has been translated into nine languages: Arabic, Dutch, Portuguese, Spanish, Korean, Vietnamese, Romanian, Turkish, and Chinese.

Several other interpretations of big history have appeared in recent years. These can be divided into those from scientists, those from historians, and those from the wisdom traditions. The astrophysicist Eric Chaisson laid out seven epochs of increasing complexity with his book for general audiences, *Epic of Evolution: Seven Ages of the Cosmos* (2006). The astronomer Russell Genet wrote an accessible, humorous account, *Humanity: The Chimpanzees Who Would Be Ants* (2007). The biochemist/anthropologist Fred Spier, the author of the pioneering *The Structure of Big History* (1996), refined his theoretical framework in his recent book, *Big History and the Future of Humanity* (2010).

From the historians, David Christian's *Maps of Time: An Introduction to Big History* (2004) remains the seminal account. It topped the charts as University of California Press's bestseller, and a new edition came out in 2011. Christian's taped lectures on big history with the Teaching Company and his TED talk in March 2011 have also reached a wide audience.

An early account of big history from the wisdom traditions appeared in 1992 as *The Universe Story*, written by the Catholic priest/cultural historian Thomas Berry and the cosmologist Brian Swimme. This tradition has been carried forward by Swimme and Mary Evelyn Tucker, professor of comparative religion and environmental studies at Yale University, resulting in a fifty-five-minute movie aimed at general audiences, *Journey of the Universe* (2011). This film, accompanied by educational materials, is being shown frequently on television and has been appealing to widespread audiences in the United States and Canada.

For those who want the story in verse, Betty-Ann Kissilove has published *Great Ball of Fire: A Poetic Telling of the Universe Story* (2010), and James Lu Dunbar has created not only verse but cartoon illustrations of the whole story: *Bang! The Universe Verse, Book 1* (2009) and *It's Alive! The Universe Verse, Book 2* (2011).

Now that big history has appeared in various versions and voices, we can begin to ask: does the big history story have an underlying theme that can be agreed upon? In my book, I steered away from proclaiming an underlying theme or pattern—too many people have been mistaken in the past for me to be that bold. Instead, I focused on the relationship of humans to the environment as the common thread in my narrative.

The field of big history is new enough that its authors are still feeling their way toward identifying an underlying theme or pattern. However, some consensus seems to be emerging that a pattern does underlie the cosmic narrative, the pattern of increasing complexity. Swimme and Tucker, in the movie *Journey of the Universe*, call this the creativity of the universe—the unforeseen, completely new properties that keep emerging as the universe evolves.

Scientists and historians do not usually use the word "creativity," but they mostly agree that increasing complexity has emerged steadily over the existence of the universe. Eric Chaisson and Fred Spier have done the most to clarify this idea. They describe the nature of complexity as diverse components bound into larger structures, which display emergent properties—that is, they give rise to new and larger structures. Energy must flow through the structures to bind the components. This energy flow can be estimated as the amount of energy (in ergs) flowing through a given mass (in grams) in a given period of time (in seconds). Chaisson, for example, estimates the energy flow, which he terms the "free energy rate density," in a galaxy as 1, in stars as 2, planets 75, plants 900, animals 20,000, a human brain 150,000, and modern society 500,000 (*Cosmic Evolution: The Rise of Complexity in Nature* [Cambridge, MA: Harvard University Press, 2001], p. 140).

Despite the frequent use in big history accounts of the pattern of increasing complexity, I have chosen to leave this book as written, without a clear argument for pattern. I am comfortable with arguing that complexity has increased so far—as I do in a forthcoming university textbook, tentatively titled *Big History: Between Nothing and Everything*, by David Christian, Craig Benjamin, and myself—yet I prefer in this book to emphasize that our connection to our environment is what matters most.

Meanwhile, proponents of big history have been making their own history. At the University of Southern Maine, biologist/historian/archaeologist Barry Rodrigue introduced a course on big history and created, with Daniel Stasko, a Listserv of professors interested in big history. At the University of California–Berkeley, the world-renowned geologist Walter Alvarez initiated a course in big history and eventually invited a small group of big historians to spend a week in Coldigioco, Italy, to learn more geology, which Alvarez rightly found underdeveloped in many accounts of big history. Out of the meeting in Coldigioco came the formation of a scholarly organization to promote teaching and research in big history—the International Big History Association (IBHA). It now has headquarters at Grand Valley State University in Grand Rapids, Michigan, and will hold its first conference in August 2012 in Michigan. (See www.ibhanet.org.)

At the college/university level, some forty to fifty courses dealing with big history are currently being taught. Out of Alvarez's course at UC Berkeley came a project called ChronoZoom, which was picked up by Microsoft and develops an interactive timeline of big history designed to serve university teachers and students. (See www.chronozoomtimescale.org.) The faculty at the Stone Age Institute at Indiana University has undertaken many big history initiatives. Among them are several versions of a big history course, a museum of universe history, and the Stone Age Institute Band. (See www.bigbangtowww.org.)

Faculty members at my own university, Dominican University of California, have created an extensive program in big history. The faculty voted that, beginning in fall 2010, all entering freshmen would be required to take a sequence of two big history courses, the first a core big history survey, followed by a choice of one discipline to examine through the lens of big history. In their second year, students can choose additional courses on the theme of "shaping the future," a way to carry big history forward in their thinking and plan their lives around its insights. This program was developed under the leadership of Mojgan Behmand, a professor of English, and includes the collaboration of creative faculty members across the disciplines in designing the courses. (See http://www.dominican.edu/academics/big-history.)

The sooner young people encounter the clarifications of big history, the better. College seems a bit late, as Bill Gates realized in 2010 when he watched David Christian's big history lectures on the Teaching Company tapes. Gates and Christian have since collaborated on the Big History Project, which is producing an online high school big history course aimed at ninth grade students, whose curricula still allows for some electives. This course has been piloted with much enthusiasm and success in 2011 and 2012 by high school teachers in the United States and Australia and is expected to be available online for free in 2013. (See www.bighistory project.com.)

Yet ahead of us all in cosmic thinking was the Italian Catholic educator Maria Montessori (1870–1952). She and her son, Mario Montessori, worked out their idea of cosmic education for six- to twelve-year-olds while they were in India under house arrest by the British during World War II. Today Montessori educators in an estimated twenty thousand schools worldwide are telling children about their place in the universe using books and materials created by Montessori educators and also those by Jennifer Morgan, including *Born with a Bang, From Lava to Life*, and *Mammals Who Morph*.

Giving young children the cosmic story in ways that they can understand helps them grasp the reality of who they are and where they came from. It also helps them build a solid foundation of scientific knowledge. Since many university freshman currently seem to be uncomfortable with science in general, it appears that an early exposure to scientific material is crucial.

Even the human part of the big history story is not easy for city dwellers to grasp. For 99 percent of our history—some two hundred thousand years—most humans were nomadic hunters and gatherers, something completely out of the experience of today's city people. Even the recent ten thousand years, when most humans were farmers, is beyond the experience of today's urban population. By 2010 more than 50 percent of the world's population was living in cities, rendering the natural world and human evolution increasingly abstract for more than half of us.

My own life experiences gave me a concrete basis from which I could imagine human evolution on the large scale. My parents, both of whom grew up in rural southern Wisconsin, longed for a wilderness experience. In 1949, when I was eleven years old, they arranged to fill this longing by taking our family to Lake of the Woods, Canada, where we lived for a month in a log cabin on an island accessible only by portage and boat. Every day we went fishing, guided by Walter Redsky, the chief of the local First Nation tribe. We ate our catch for lunch. We swam, canoed, and ate wild blueberries that we picked on granite outcroppings. It was as close to foraging in the wilderness as we could come, and it left an indelible impression on me.

My experience of agricultural life occurred during summer trips to my maternal grandparents' farm in southern Wisconsin. They kept a herd of Holstein dairy cows to sell their milk and grew peas and red beets for the local cannery. Grandma fed many city cousins, plus the neighboring farmers when they came to help thresh the oats, from her large garden and orchard. As children, we drove the cows to pasture and back, carried water to the men hoeing in the fields, and rang the bell for them to come for lunch. Later we rode on top of wagons loaded with bales of hay and learned to drive the tractor. It was an idyll for carefree children, though my grandparents could not escape milking the cows twice a day.

Before writing big history I had another crucial experience, this one among nomadic pastoralists in the mountains of Central Asia. As we trekked to the base camps of K2 and later Khan Tengri, I was able to sleep in a yurt and mingle with people taking their flocks to high summer pastures.

Without these experiences, would I be able to imagine humanity emerging from hunting-gathering into agriculture and civilization? I'll never know, but this feat of imagination must be more difficult for young people without my experiences. Many of them have access to previously unimaginable information, but they have so little time for reflection and so little immersion in the nonhuman natural world that I cannot help but wonder what kind of adults they will become. What powers of imagination will they have? What kind of story will they tell about the human journey?

Turning now to the reality that we humans currently find ourselves in, what has been the environmental news of the last five years? Thankfully, this news is now making headlines; in the years prior to 2007, one could find only tiny articles about the environment used as fillers on inside pages of newspapers. Awareness of the environment, and in particular of climate change, has increased enormously, assisted by the Al Gore film *An Inconvenient Truth* (2006).

Much has been learned about our environment in the past five years, and the news has not been encouraging. The human impact on this planet has grown to such proportions that since 2008 geologists are seriously considering proclaiming the end of the Holocene epoch and the beginning of a new geologic epoch, the Anthropocene: one of human domination.

Perhaps the aspect of the environment receiving the most coverage recently is that of global warming—or climate change, since it does not always include warming. Plants and animals are moving north; glaciers are melting, storms and droughts are increasing in strength and severity. Re-

cent polls in the United States suggest that only about 60 percent of its people believe that global warming is occurring (one poll indicated that 78 percent of registered Democrats believe in global warming versus 34 percent
of those belonging to the Tea Party). Many of those who believe global
warming is occurring do not attribute the cause to human activity but
rather to planetary trends.

Climate scientists make a range of predictions, of course, but they
agree that the data demonstrate that change is occurring more rapidly
than predicted. Scientists also agree that humans need to reduce our CO_2
emissions rapidly to avoid a level of climate change disastrous to civilization as we know it, reductions on the scale of 4–5 percent a year. Instead, in
2011 the two largest emitters, China and the United States, increased their
CO_2 emissions by 10 percent and 4 percent, respectively, with a global average increase of 6 percent. The measurement of CO_2 in the atmosphere is
now at about 390 parts per milligram (ppm), while many scientists believe
that it must be reduced to at least 350 ppm.

(It is difficult enough to write these figures. One can easily understand
why many people deny that any of this is occurring.)

The increase of CO_2 in the atmosphere also affects the oceans, which
become more acidic as they absorb more CO_2. The increasing acidity
threatens all creatures that form calcium carbonate shells and plates—
marine algae and plankton, coral, snails, crabs, lobsters, shrimp. The chem-
istry of the oceans is further affected by the extra nitrogen in fertilizer
runoff and by plastics beaten into atom size by the action of the water and
absorbed into living creatures. The combination of these actions, plus over-
fishing, means that the health of oceans and marine life everywhere has declined in the past five years.

Biologists are alarmed by the seemingly increasing rate of extinction of
species. Much progress has been made in identifying living species, but still
only a small fraction of all living species is known; hence, the rate of extinction is hard to calculate. A natural background rate of extinction accompa-
nies the usual process of evolution, but biologists believe that the current
rate exceeds the usual rate by somewhere between a hundred and a thousand times. Many experts believe that we are now witnessing a global mass
extinction and that before it is over it will rank as the sixth major extinction in the history of our planet.

To counter this dismal information, there is some hopeful news. Much
new investment has been made in the development of sustainable technologies, and much political action has been undertaken around the world.
For example, my home state of California would be the twelfth-largest

emitter in the world if seen as a separate country. In 2006 California legislators passed, and Governor Arnold Schwarzenegger signed, legislation to create a green economy based on solar and bio energy and nuclear power and to reduce CO_2 emissions by 2020 to 1990 levels. A proposition to suspend this legislation was defeated by a 61 percent majority vote in 2010.

In order to make a quick global transition away from burning coal, gas, and oil for energy, some fossil fuel companies and their stockholders may have to lose a lot of money, which they have invested in facilities that they expect to pay off with future profits. Most of them will not volunteer to do this; they must be defeated by political battles in which the citizens of the world mobilize to defend their interests. Scientists advise us that we likely have only about ten years in which to change our destructive behavior and implement sustainable techniques before our climate will be irreversibly out of control. Trusting to the demonstrated capacity for innovation that humans have shown in our history, we can hope that such sustainable techniques will emerge. But the political determination must also emerge. For that to happen, I hope my account of humans in their cosmic and environmental setting will be of some small assistance.

I cannot close without expressing my deep gratitude to the amazing people I have met along the way of big history. This new field has attracted big-minded, bighearted people from around the world. In addition to those already mentioned, I am particularly grateful to my colleagues at Dominican University of California, who have developed big history in ways I never dreamed of. The highly skilled leadership of Mojgan Behmand has guided everyone. Others include: Martin Anderson, Arturo Arrieta, Tom Burke, Jaime Castner, Heidi Chretien, James Cunningham, Lindsey Dean, Judy Halebsky, Dan May, Phil Novak, Rich Simon, Lynn Sondag, Harlan Stelmach, Cynthia Taylor, Neal Wolfe, and Julia Van der Ryn. Through their creativity they have used the perspective of big history to transform their studies, their classrooms, and their institution.

My deepest gratitude goes to David Christian for receiving this book warmly, and to him and Craig Benjamin for inviting me to collaborate with them in writing a university textbook of big history. That book is not yet published, but writing it enabled me to go through the whole story again with the stimulation of supportive, provocative colleagues, making for a grandly enjoyable and meaningful experience.

Preface and Acknowledgments

Big History presents the scientific creation story, from the big bang to the present, told in succinct, understandable language. In this book I have woven many disciplines of human knowledge together into a single, seamless narrative.

History as a discipline traditionally begins with written records from about 5,500 years ago. Here I have extended "history" to the limits of what is currently knowable by scientific methods, using whatever data and evidence are available, and not limited to written documents. History is part of the scientific undertaking, and there is no sound reason why the uncovered story should be cut into two segments, one labeled "science" and the other "history."

We need to extend our story backward, for the five thousand years of recorded history tells only a millionth of the lifetime of Earth. To understand the kind of world we live in and the kind of creature we are, we must look beyond the written record.

Nor, do I believe, is there any sound reason to label one part "religion" and the other "science." Within the last fifty years the scientific community has established a verifiable, and largely verified, account of the origins of our universe—of where we came from, how we got here, and where we may be going. This is a creation story for our time—for a world built on the discoveries of modern science, a world of jet travel, heart transplants, and the worldwide Internet. This world will not last forever, but while it does, this is our story.

We are now able to think in scientific terms about the timescales of the universe we are part of—its beginning, middle, and end—and thus with current thinking we are able to put the story of our planet into its larger context. The power of our thinking and imagination has led us onto a scale that brings cold comfort to many. For many others, myself included, our significance as humans increases rather than decreases against the scale of the universe. I try to report the facts as they are known now, without attempting to discuss or resolve our contrary human responses to them, well aware that the facts are ever changing.

What, you may well ask, is my overall approach in telling this tale? A story must be told with some plot, some theme. Every author who writes a big history does so with some unique emphasis, some unique voice.

I try to stick to the information and theories well-accepted in the scientific community, staying as unopinionated as is humanly possible. I am telling a story, not making an argument. As a historian, I give more space to human history than a geologist or biologist might. I strive to keep the story as simple as possible, without violating too much the unending complexity and contradictions of history. I put in lots of what I consider most basic: climate, food, sex, trade, religion, other ideas, and empires/cultures.

Some subtle emphasis, of course, has to recur to keep any story from dissolving into a muddle. In this book, that underlying theme is the impact of human activities on the planet, as well as the planet's impact on people. When I combined the story of the planet and the people on it, I found that the actions people have taken to keep our offspring increasing have put the planetary environment and its life-forms in grave jeopardy. Summed up in one phrase, this story portrays the "increase of people" rather than the "ascent of man."

This theme emerged as I wrote the story rather than the other way around. Clearly, my mind was involved in the telling, so perhaps it is more accurate to say that this theme is what I noticed recurring as I tried to tell the entire human story as compactly as I could without truncating it to begin with farming. Only the longer time frames reveal what humans have wrought; I was partially, but not fully, aware of it until my story was told.

The person who most spurred me to put the whole story together is David Christian, now professor of history at San Diego State University in California. From 1975 to 2000 Christian taught Russian and European history at Macquarie University in Sydney, Australia. In 1989 he launched a course there that he called, in jest, "big history," as a way to show colleagues what he thought an introductory course in history ought to look like. The semester-long course started from the beginning—that is, from the origin of our universe. Christian began with lectures on time and creation myths, followed by colleagues invited from other departments lecturing on their specialties. In an article in the *Journal of World History*, Christian described his experience with this course. Reading that article turned my thinking in a new direction. "Big history" has become the current term designating this endeavor, and in 2004 Christian published his majestic overview of the story and the technical issues involved in big history: *Maps of Time: An Introduction to Big History*, which I resisted reading until I had finished my first draft of this book.

An earlier pioneer in big history, before it had a name, was Clive Ponting, at University College in Swansea, England. His account, which I cherish, is *The Green History of the World: The Environment and the Collapse of Civilizations*. Ponting did not start with the big bang, but he did devote a chapter to "The Foundations of History," in which he described the influences of large-scale geologic and astronomical forces over long periods of time.

Because I had to laugh a lot to get started on this project, I treasure two other early big histories: Larry Gonick, *The Cartoon History of the Universe: From the Big Bang to Alexander the Great* and Eric Schulman, *A Briefer History of Time: From the Big Bang to the Big Mac*.

Big history, defined as history from the big bang to now, is still a tiny subfield of the history subspecialty of world history, which itself has had its own journal only since spring 1990. Big history has no journal yet and only a handful of practitioners worldwide who actually teach overt big history courses at universities. Other professors may be tacking on universe and planetary history as an introduction to world history or to world religions. How have I, as one of these early practitioners of big history, been able to get around the academic obstacles and disciplines to teach it and to write this book?

To answer, I must begin with my mother, Louise Bast Stokes, who set me on my course by her wide-ranging intellectual interests: astronomy through geology and biology to world religions. As a middle-school biology teacher in the early 1930s, she accepted evolution as the underlying principle of life and pointed out the living world around me through this lens. Thus, "big history" is a natural way of thinking for me, a gift from my mother.

Growing up in a small town in western Kentucky, I had a bicultural experience right inside the USA. My parents had grown to adulthood in southern Wisconsin, but after marriage in 1935 they lived in eastern Kentucky, where Dad built roads through the mountains. With my impending birth (1938), they settled in western Kentucky, in Madisonville, where Dad and his partners bought and operated a small strip coal mine. My parents were immigrants in a strange, Southern culture into which Dad assimilated as thoroughly as he could, while my mother clung to her native Wisconsin customs and values. Multiple perspectives were thus built in for me, along with a love of storytelling, a gift from my father.

Identifying with my mother, I never felt part of the South, yet I stayed there for my college years at Duke University in Durham, North Carolina. I earned a Master of Arts in Teaching degree at Johns Hopkins University and started teaching world history to high school students in Baltimore, Maryland. With the encouragement of my professors at Hopkins, and supported by fellowships from the Woodrow Wilson Foundation and the American Association of University Women, I finished a Ph.D. at Hopkins in 1964 in the history of education, with a dissertation about the first four Americans to study at a German university in the early nineteenth century.

My first son was born three months after I finished my Ph.D. and my second two years later in the northeastern Brazilian city of Fortaleza, where my first husband was serving as a Peace Corps physician. Living in Brazil for two years exploded my cultural assumptions and opened me to world history. My first published writing was about the great Brazilian educator, Paulo Freire, who fled Recife in 1964, just a year before we lived there.

After Brazil, I stayed home with my sons in Baltimore, and in 1969 we moved to Berkeley to begin a new life there in a much more open culture than any we had experienced before—one oriented to the Pacific as well as to New York City and Europe. Momentous shifts were occurring—multiculturalism, the *Whole Earth Catalog* started by Stuart Brand in 1968, and, in the same year, the first precious photos of our fragile planet floating in space.

When I was ready for a full-time academic job (1981), I found it in the school of education at Dominican University of California, then Dominican College, directing the single-subject teaching program. I subscribed to the first and all subsequent issues of the *Journal of World History* and helped to bring to my campus an in-service program for teachers called Global Education Marin, to help them globalize their curriculum. This program later became part of a statewide effort called the International Studies Program, out of Stanford University. In this way I kept up with developments in world history and found my way to Christian's article.

With my new orientation toward big history, I sought ways to express my ideas. In the spring of 1992, I taught a course in the history department called "Columbus and the World Around Him," and in 1993 I taught an undergraduate world history course for those preparing to become elementary teachers. I started this course with my own account of the big bang and the evolution of life, used Ponting's book as a text, and asked students to construct timelines from the big bang to now. Students received the course with great enthusiasm; it didn't daunt them, only me.

Having returned to full-time work in the school of education, I proposed, when it came my turn for a sabbatical, to write a history of the world. Half the committee thought that a terrific idea, while the other half laughed aloud. So, to secure my sabbatical, I temporarily abandoned world history and wrote instead *Refusing Racism: White Allies and the Struggle for Civil Rights*.

After my retirement from full-time teaching, all I wanted to do, after a brief rest, was to put together this story. I began writing in late September 2002, following the death of my mother, and finished my first draft in December 2004. I used files of articles from the *New York Review of Books* that I had been saving for twenty years; thank you, Bob Silvers and Barbara Epstein. I read the dazzling work of contemporary scholars; thank you, Timothy Ferris, Lyn Margulis, Stephen Pinker, Jared Diamond, J.R. and William H. McNeill, and David Christian.

To test my ideas with students, I returned to part-time teaching in the history department. I continued my course with prospective elementary teachers and also created a colloquium, or set of three courses from different departments on an interdisciplinary theme, which we called "The Universe Story." I am grateful to the traditions of Dominican that feature this kind of interdisciplinary teaching. Our colloquium consisted of my course, "Whole Earth History"; Jim Cunningham's from the science department, called "Life on Earth"; and Phil Novak's from the religion/philosophy department, called "The World's Religions." Again, students responded enthusiastically, hardly seeming to notice that we were doing anything unusual. I am greatly indebted to these colleagues for their boldness and confidence in undertaking this project, leaping across all academic boundaries without a quiver.

More than ever in my writing, my colleagues, family, and friends participated in this book. My chair in education, Barry Kaufman, and my colleagues in the history department, especially Sr. Patricia Dougherty, O.P., and Martin Anderson, offered frequent support. My colleagues in Global Education Marin—Nancy van Ravenswaay, Alice Bartholomew, and Ron Herring—kept me going in the right direction over many years. My sister, Susan Hill, and her son, Ian Hill, asked eagerly for each new chapter, as if they couldn't wait. My stepdaughter, Deborah Robbins, who teaches world history at University High in Los Angeles, could discuss any issue and lead

xxiv PREFACE AND ACKNOWLEDGMENTS

me to new ones. One son, Ivor, sent me leads to books and articles, while the other, Erik, kept me in touch with the importance of good food. My aunt Jean in El Salvador, and her husband, Jorge Bustamante, were always an inspiration. Friends around the world each contributed to my understanding.

I am grateful to numerous early readers. Amit Sengupta, physics and math professor at Dominican, checked my first chapter, while Jim Cunningham, professor of biology, checked my second. My history colleague, Martin Anderson, saved me from a number of gaffes. My religion/philosophy colleague, Phil Novak, saw my whole picture quickly and gave me confidence in it, despite its materialist assumptions. World historians John Mears and Kevin Reilly provided expert advice. David Christian helped in immeasurable ways. Other readers each made important contributions: Jim Ream, Chester Bowles, Margo Galt, Katie Berry, Marlene Griffith, Joan Lindop, Philip Robbins, Susan Rounds, and Bill Varner. My husband, Jack Robbins, read every draft; his love and support made my work possible.

I feel special gratitude to the staff of The New Press, especially Marc Favreau, Melissa Richards, and Maury Botton, for executing this project with great enthusiasm and professionalism.

The mistakes and misjudgments that remain are my own.

BIG HISTORY

PART I

The Depths of Time and Space

1

Expanding into Universe

(13.7 Billion–4.6 Billion Years Ago)

We are all whirling about in space on a small planet, bathed for part of each day in the light and warmth of a nearby star we call the sun. We are traveling 12 million miles a day around the center of the Milky Way galaxy, which is whirling in a universe of more than 100 billion galaxies, each home to 100 billion stars (Fig. 1.1).

This universe in which we whirl began as a single point 13.7 billion years ago; it has been expanding ever since, with its temperature steadily decreasing. Our universe has at least four dimensions, three of space and one of time, meaning that time and space are interconnected. Just now the size of our observable universe is roughly 13.7 billion light-years on each of three dimensions by 13.7 billion years on the dimension of time, increasing as I write and you read.

Ever since human beings developed, they have been looking at points of light in the nighttime sky with awe and respect, learning what they could from direct observations and using this knowledge to make predictions, to travel on land, and to navigate by sea. Without specialized instruments, however, people could not detect much about the origin of our immense universe and the nature of matter, because the scale of the universe and of matter is so different from that of everyday life. By the late twentieth century, scientists had invented instruments that could begin to view the macroscopic heavens and the microscopic domain. Knowledge about these worlds has recently expanded exponentially. Now everyone can understand the amazing universe that is our home—if we use our imaginations and absorb the photographic images and diagrams that are currently available.[1]

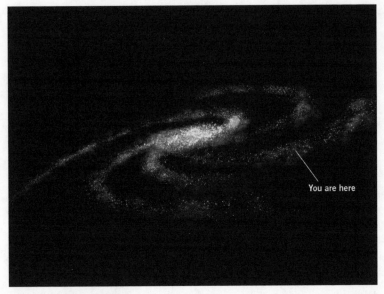

You are here

1.1   The Milky Way Galaxy
This drawing shows that our solar system orbits about halfway out to the edge of our galaxy.

Fog and Transparency

It all began with an inconceivable event: the big bang. (This name was given by the British astrophysicist Fred Hoyle on a BBC radio broadcast in 1952.)[2] The universe erupted from a single point, perhaps the size of an atom, in which all known matter and energy and space and time were squeezed together in unimaginable density. Compressed space unfurled like a tidal wave, expanding in all directions and cooling, carrying along matter and energy to this very day. The power in this initial expansion was sufficient to fling a hundred billion galaxies for 13.7 billion years and counting. The billowing universe was under way.

Where did this eruption take place? Everywhere, including where each of us is right now. In the beginning all the locations that we see as separate were the same location.

Initially the universe was composed of "cosmic plasma," a homogeneous substance so hot that it had no known structure at all. Matter and energy are interchangeable at temperatures of many trillion degrees; no one knows what energy is, but matter is energy at rest. As the universe cooled, the smallest constituents of matter that we know about, called quarks, began to clump together in groups of three, forming both protons

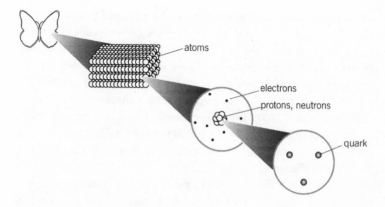

1.2 The Constituents of Matter
Matter is composed of atoms, each of which is composed of electrons circling a nucleus containing protons and neutrons, both of which are made of quarks. Whether quarks are composed of something smaller is currently unknown.

and neutrons (Fig. 1.2). This took place at about one hundred thousandths of a second after the big bang, when the temperature had cooled to about a million times hotter than the sun's interior. A hundredth of a second later, these protons and neutrons began hanging together to form what would later become the nuclei of the two lightest elements, hydrogen and helium.

Before one second had elapsed the four fundamental forces that govern matter had come into being: gravitational force, electromagnetic force, the strong nuclear force, and the weak nuclear force. Gravitational force, or gravity, is the weakest of the four forces. It was described by Newton's theory of gravity and by Einstein's general theory of relativity, but it still cannot be defined. Electromagnetic force is a union of the electric and the magnetic forces. The strong nuclear force, the strongest of the four, is responsible for keeping quarks locked inside of protons and neutrons and for keeping protons and neutrons crammed inside of atomic nuclei. The weak nuclear force mediates the decay (or disintegration of the atomic nuclei) of radioactive elements. Scientists believe that all four forces must be aspects of one force, but they have not yet been able to create a unifying theory.

These four forces work in perfect balance to allow the universe to exist and expand at a sustainable rate. If the gravitational force were a tiny bit stronger, all matter would likely implode in on itself. If gravity were slightly weaker, stars could not form. If the temperature of the universe had dropped more slowly, the protons and neutrons might not have stopped at helium and lithium but continued to bond until they formed iron, too heavy to form galaxies and stars. The exquisite balance provided by the four forces seems to be the only way in which the universe can maintain itself. Scientists wonder if perhaps many other universes came into existence but vanished before this one survived. The newborn universe evolved with phenomenal speed, setting in place in a tiny fraction of a second the fundamental properties that have remained stable since.

During about 300,000 years of expanding and cooling, the wildly streaming electrons, negatively charged, slowed down. The atomic nuclei, protons and neutrons, were positively charged. When the electrons had slowed down sufficiently, the nuclei could attract them by their electric charge and form the first electrically neutral atoms: hydrogen (H) and helium (He), the lightest elements, the first matter. Hydrogen consists of one proton and one electron; helium consists of two protons and two electrons.

This became a pivotal moment in the story of the universe. Before the formation of stable atoms, the universe was filled with so many zigzagging particles, some negative, some positive, that light (consisting of subatomic particles called photons) could not move through the bath of charged particles. This was so because photons interact with electrically charged particles and are either deflected or absorbed. If anyone had been there to see it, the universe would have appeared as a dense fog or a blinding snowstorm.

As soon as atoms formed, binding the negative electrons and positive protons together, the photons of light could travel freely. The dense fog of radiation lifted. Matter had formed, and the universe became transparent. Its full expanse came into view—if anyone had been there to see it—consisting mostly of vast empty space filled with huge clouds of hydrogen and helium with immense amounts of energy pouring through them.

Today we can see some of the photons left from the big bang—as "snow" on our television screens. To do so we must disconnect the cable feed and tune to a channel the set does not receive. About 1 percent of the "snow" we see is residual light/heat left from the big bang that forms a cosmic sea of background microwave radiation.[3] If our eyes were sensitive to microwaves, which they are not, we would see a diffuse glow in the world around us.

By using radio equipment, scientists have documented the background microwave radiation. By the 1950s and 1960s physicists realized, from what they already knew about the universe, that the present universe should be filled with primordial photons, cooled over 13.5 billion years to a few degrees above absolute zero. In the spring of 1965 two radio astronomers, Arno A. Penzias and Robert W. Wilson, working for Bell Laboratories in

New Jersey, accidentally detected this afterglow as a background hissing noise while they were testing a new microwave antenna to be used with communication satellites. In 1989 NASA sent up the Cosmic Background Explorer (COBE) satellite, which collected information that confirmed with high precision that there are about 400 million photons in every cubic meter of the universe—an invisible cosmic sea of microwave radiation, at 3 degrees above absolute, just as predicted by the theory of the big bang.

In 2002 NASA sent a sixteen-foot probe called the Wilkinson Microwave Anisotropy Probe, or WMAP, a million miles out from Earth. For a year WMAP took time exposures of the entire sky, showing in high resolution the map of the cosmic background radiation (CBR) from 380,000 years after the big bang and confirming again the big bang account of the universe.

Fortunately for astronomers, on the scale of the universe, distance is a time machine. The farther away something is, the younger we see it; this is because the more distant something is, the longer its radiation takes to reach us. We can never see the universe as it is today, only as it once was, because it takes millions and billions of years for the light of distant galaxies and stars, traveling at nearly 6 trillion miles a year, to reach us. Hence, we can see far back into the past. By picking up microwave radiation, we can "see" back nearly to the beginning of the universe (Fig. 1.3).

Think of it this way. The light from our nearest star, the sun, takes eight minutes and twenty seconds to reach us. Light from Jupiter takes about thirty-five minutes when it is closest to us, about an hour when it is farthest away in its orbit. The light of the brightest star in the night sky, Sirius, takes 8.6 years to reach us. (The distance the light travels is 8.6 light-years, or 50.5 trillion miles). The light from stars we can see without optical aid takes from four years to 4,000 years to reach us. If we should see a star exploding 3,000 light-years away, then that explosion occurred 3,000 years ago—the time it takes for the light to reach us.

Twinkling Galaxies

As described earlier, the universe became transparent some 300,000 years after the big bang. Immense clouds of hydrogen and helium drifted until these clouds broke into about a trillion separate clouds, each with its own dynamics, each escaping from the universe's expansion in that the diameter of each cloud remained the same while the space between the clouds increased.

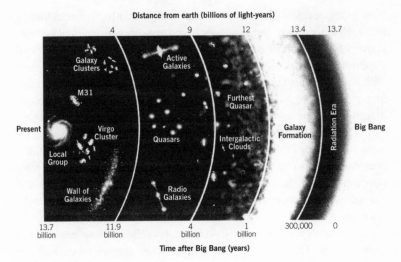

Distance from earth (billions of light-years)

4   9   12   13.4   13.7

Galaxy Clusters   Active Galaxies

M31   Furthest Quasar

Radiation Era

Present   Virgo Cluster   Galaxy Formation   Big Bang

Quasars   Intergalactic Clouds

Local Group

Wall of Galaxies   Radio Galaxies

13.7   11.9   4   1   300,000   0
billion   billion   billion   billion

Time after Big Bang (years)

1.3 Our View of the Universe

From our position in the Milky Way galaxy—one of the galaxies in the Local Group—we see the universe in the distant past, because the light from remote galaxies takes billions of years to reach us. In this distant past the universe was smaller, and galaxies collided more often. Quasars are very distant objects thought to be the nuclei of younger galaxies, possibly in collision.

As the universe cooled and calmed down, each separate cloud of hydrogen and helium became a separate galaxy of stars joined by gravity. This happened as the atoms of hydrogen and helium collided with each other. As they collided, the friction created temperatures so high that the atoms were stripped of their electrons. The hydrogen nuclei started to fuse, forming helium ions. These fusion reactions released a huge amount of heat/energy, according to Einstein's equation $E = mc^2$, in which the loss of a tiny bit of mass results in energy multiplied by the speed of light squared. As the hydrogen begins to burn, millions of tons of matter are transformed into energy each second, and a star is born. The earliest stars formed only about 200,000 years after the big bang.

The universe is filled with an enormous range of objects as measured by their mass. The largest objects are stars, which produce their own energy. The largest stars are up to twenty times more massive than the star that is our sun. The smallest objects in the universe are dust particles visible only under a microscope and which rain down into the Earth's atmosphere at the rate of a hundred tons a day. The silt in the eaves of any house probably contains a minute amount of interstellar material. Planets are middle-range objects; their mass is not sufficient to produce their own energy through hydrogen-fusion reactions.

Stars come in a vast range of sizes and densities, and they evolve over time from one type to another. Most of the stars nearest us are red stars, but the one we know best, the sun, is a stable yellow star burning hydrogen, called hydrogen fusion as described earlier. When its hydrogen is used up, in about 5 billion years, our sun will switch to burning helium, called helium fusion. Since helium fusion is a hotter process with a greater energy output, the pressure from the extra energy will expand the sun until it becomes what is called a red giant. When the helium fuel is used up, the red giant will collapse to a white dwarf. Then it will slowly cool until it becomes a cinder called a black dwarf, about the size of Earth and 200,000 times its mass. No black dwarf has yet been found because the universe is not old enough for any to have completed the slow process of cooling down.

Some yellow stars, the ones that are larger than our sun at their inception, become larger red giants than our sun will. When their red-giant stage is over, they do not shrink into white dwarfs. In them heavier elements are created and burned: carbon, nitrogen, oxygen, magnesium, and finally iron. But iron cannot be used as a stellar fuel. Energy production stops and gravity takes over. The star's core implodes and triggers an immense explosion of the outer layers that blasts most of the star to smithereens. Only the core survives as a white dwarf, a neutron star (tiny and incredibly dense), or a black hole, which is an object so dense that light cannot escape its gravitational field. This explosive self-annihilation of a star is called a supernova; only stars at least six times more massive than our sun can become supernovas.

These supernovas play an immense role in the creativity of the universe. They are the cosmic furnaces out of which new elements are formed and, and we have seen, they initiate the formation of black holes. When a star of more than ten times the mass of our sun explodes, the imploding core that is left may be larger than four times the mass of the sun. If it is, then gravity is so immense that all the matter disappears and a black hole remains, leaving only a gravitational field so strong that it prevents light from escaping. No one knows where the matter goes. The center of a black hole is called a singularity; a black hole created by a star of ten solar masses has a diameter of only forty miles. Around the singularity is a field of gravitational force so powerful that anything that enters the field disappears into the hole.

Astronomers suspect that massive black holes exist at the core of most galaxies, as one seems to at the center of our Milky Way galaxy. Our black hole, a few million solar masses, is called SgA because it appears to lie in the southern hemisphere constellation Sagitarius. Scientists, working for over ten years at the Very Large Telescope in Chile's Atacama Desert, confirmed in 2002 the presence of SgA.

Enormous supernovas become black holes. Smaller ones, those between three and six solar masses, explode outward rather than implode inward. In their burning cores hydrogen is burned into helium, then helium to carbon; nuclei are fused into ever larger nuclei, like oxygen, calcium, and on through the periodic table of elements. At some point an explosion occurs, spewing most of the star back into space as gas, but now containing complex, life-supporting atoms, not merely hydrogen and helium.

Only supernovas can create elements higher than iron. Gradually, over roughly 9 billion years, all the elements of the periodic table were built up in this way. Every scrap of gold on our planet originated in giant stars that exploded before the sun was born. The gold in the ring on your finger has to be more than 4.5 billion years old. Thus explosions of stars created the elements that make life on Earth possible. We quite literally are made of stardust.

Coming back to our story, several hundred thousand years after the big bang, galaxies consolidated as density waves moved though space, shocking the clouds of hydrogen and helium into star formation. Space began to twinkle, with billions of stars flowing in spidery filaments of whirling spirals. Most galaxies took the shape of spirals, but in the early universe matter was crowded, and galaxies often bumped into one another. When they did, the large one absorbed the smaller, but the large one could never recover its spiral shape. Instead, it became a sphere or an ellipse (oval), called an elliptical galaxy. Elliptical galaxies do not produce new stars, since density waves do not move through them to shock the clouds of gases into forming new stars. Our Milky Way galaxy is a perfect spiral, the lucky accident of being in a noncongested area of the early universe about 12 billion years ago.

For some 9 billion years, the first two-thirds of its lifetime thus far, the universe consisted of unimaginable celestial fireworks. Galaxies wheeled and collided. Density waves surged through galaxies, causing new stars to form. Supernovas exploded, scattering new gaseous elements ready to be shocked into new stars by other supernovas or imploding into black holes, losing their matter to who knows where. All the while, space was expanding and the temperature cooling. The universe was a sparkling dance of death and resurrection, ruin and elegance, overwhelming violence and destruction cycled with dazzling beauty and creativity.

The Sun/El Sol/Helios/Die Sonne

About 4.6 billion years ago, in the Milky Way galaxy, a supernova exploded, and a new star—our sun—emerged from the debris. We know this because moon rocks and meteorites, all originating in that supernova, consistently date about 4.56 billion years ago.

This sun was an average-sized star, distinguished by not having a companion star (about two-thirds of the stars in our section of the Milky Way are multiple-star systems). The sun is located two-fifths of the way out on one of the spiral arms, about 30,000 light years from the center of the Milky Way. It takes about 225 to 250 million years to circle around the center of the galaxy in an elliptical, or oval, orbit, traveling about 200,000 miles a day. Accompanied by its system of planets and other bodies, the sun has orbited the center of the Milky Way about twenty times since its origin. Its size indicates it will burn about 10 billion years; it has now burned for about 4.6 billion of those years.

Around our early sun spun a disk of leftover materials—nebulous dust and gases of many elements created by our exploding supernova. As all these gaseous elements collided, they formed small grains whose instabilities shaped the disk into bands. As centers of concentration developed in these bands, the planets emerged, with the sun's gravity making the inner four (Mercury, Venus, Earth, and Mars) heavier and rockier, while the outer ones (Jupiter, Saturn, Uranus, and Neptune) are lighter and more gaseous. Pluto, smaller than our moon, has been declared not large enough to be considered a planet. Jupiter, about 300 times the mass of Earth, is almost, but not quite, large enough to become a star.

(There is no practical way to draw the solar system to scale without using distances the size of city blocks. If Earth were reduced to the size of a pea, Jupiter would be over 1,000 feet away and Neptune would be over a mile away.)[4]

The planets in their earliest state were molten or gaseous. Each planet arranged itself by gravitational interaction; the heaviest elements, such as iron and nickel, sank into the core, while the lighter elements, such as hydrogen and helium, formed the outer layers. The static, gravitational order was broken by the unstable, radioactive elements. When these elements broke apart, their energy kept the planets in a boil, bringing materials up from the deep inside to the surface.

On the three smallest planets—Mercury, Venus, and Mars—all activity came to a halt within a billion years with the formation of rocks. On the four largest planets—Jupiter, Saturn, Uranus, and Neptune—the boiling gaseous activity continues today, similar to what it was at the beginning of the solar system. Only Earth has a size that produces a gravitational and electromagnetic balance, which allows a solid rock crust to form around a burning core. Only Earth has a position in respect to the sun, a mean distance of 93 million miles, that establishes a temperature range in which complex molecules can form. Within our solar system, only here on Earth does chemical activity continue in constant change.

We measure time by the amount of it that Earth takes to circle the sun, called one year. Earth spins on an axis while it circles the sun. This axis is tilted somewhat, about 23.5 degrees, so that Earth's electromagnetic poles are not perpendicular to the sun. Our tilted axis means that while Earth is on one side of the sun, one hemisphere leans toward the sun and receives more sunlight, and while Earth is on the other side of the sun, the other hemisphere does. This tilt of our axis as we spin creates the seasons here on Earth, for if we spun on a vertical axis both hemispheres would receive the same amount of sunlight all year round. (All other planets revolve on a vertical axis except Uranus, which revolves on a nearly horizontal axis.)

During its first half billion years the early Earth suffered the shock of collisions with meteors, asteroids, and planetoids. We need only look at the surface of our moon to see a rockscape with the imprints of these early collisions; the moon is so small that it quickly lost its internal heat and preserved its original surface. Earth was sufficiently large—with a core hot enough that the heat of those early impacts kept it boiling day and night—that no imprints of the collisions could form.

When Earth had cooled down enough for rocks to form on its surface, plumes of molten lava rose up from within, bringing chemicals forged in the interior to the surface, changing continually the Earth's atmosphere, composed mostly of methane, hydrogen, ammonia, and carbon. Gigantic electrical storms, with immense bolts of lightning and thunder, stirred the chemical pot. After some half billion years of gestation, Mother Earth lay poised to bring forth living molecules.

Unanswered Questions

My story thus far has been based on what scientists know about our universe, called the Standard Model, developed in the 1960s and 1970s. I have not knowingly strayed into speculation. Yet everything that we think we know needs to be viewed in the context of what we do not know. Many significant questions remain unanswered.

Even the origin of our moon is uncertain. Some say it is a piece broken off from Earth, but most believe that the moon arose when a planetoid crashed against Earth, could not escape its gravity, and went into orbit, knocking Earth off its vertical axis to the slightly tilted one that creates our seasons.

More difficult questions come to mind, such as: "Why do mathematical equations work to account for things like the trajectory of the moon and of the Andromeda galaxy?" and "What came before the big bang?" To the first question, mathematicians just shrug and joke, "God is a mathematician." It is simply amazing that we are able to understand anything about the universe, that our minds can create equations that correlate with reality. As for the second and other questions:

1. What came before the big bang?

No one knows what the initial conditions of the universe were. Some physicists believe the answers to this question lie forever beyond the grasp of the human mind and any of its theories. But theories abound. One, posed by Lee Smolin of Pennsylvania State University,[5] proposes that the initial condition of our universe may have been a black hole in some other universe. The description of a black hole seems similar to the story of the beginning of the universe, except in reverse—matter, energy, space, and time becoming more compacted until they disappear. Physicists who are considering Smolin's idea are theorizing that matter, energy, space, and time may disappear out of the fabric of our universe to reappear somewhere else as a new universe. Perhaps we live in a "multiverse" of many universes popping out of each other. This is just one of several current theoretical scenarios based on many universes.

2. How did the universe start expanding in the first place?

One likely hypothesis says that in its first instant of existence the universe inflated—that is, it expanded exponentially, at a rate far exceeding the velocity of light, repeatedly doubling its radius over equal intervals of time. This spasm was over in less than a second, and thereafter the universe settled into a steady linear expansion rate, until about 5 billion years ago when its expansion rate began to accelerate. This inflationary hypothesis helps explain several problems in big bang theory, but it has not been conclusively established.

3. How can theories dealing with vast, astronomical scales, called general relativity, and theories dealing with the microscopic properties of the universe, called quantum mechanics, be reconciled?

These two groups of theories contain contradictions, which cannot yet be resolved into one grand unifying theory of everything. Yet, when considering black holes or the universe at the moment of the big bang, physicists need to use both general relativity and quantum mechanics together. When they do, the answers to their equations often equal infinity. This indicates a problem, which can be stated simply in the following way: Quantum mechanics tells us that the universe on the microscopic scale is a chaotic, frenzied arena with everything appearing and disappearing unpredictably. In contrast, general relativity is based on the principle of a smooth spatial geometry. In practice, avoiding extremes of scale, the theories of quantum mechanics and general relativity work perfectly to predict perceivable outcomes; the random, violent undulations of the microscopic world cancel each other out to behave like smooth fabric.

Physicists feel that their knowledge must be considered incomplete until there are no contradictions and inconsistencies in their theories. In 1984 two physicists, Michael Green and John Schwarz, provided the first piece of evidence for a new unifying theory, called superstring or string theory, for short. This idea posits that the most elementary ingredients of the universe are not point particles but wriggling strands, or strings, of energy whose properties depend on their mode of vibration. These miniscule strings are so small—about 10^{-35} centimeters long—that they appear as points, even to the most powerful available equipment. This theory also posits that the universe has more than three dimensions plus time—maybe ten (or more) dimensions plus time. Theoretically, string theory provides a truly unifying theory, positing that all matter and all forces arise from one ingredient: oscillating strings of energy. Since 1984 additional pieces of evidence have fallen in place to reinforce the idea of strings, but experimental evidence to validate the theory has not yet been found.

4. Ever since scientists began, in the 1960s and 1970s, to feel certain that the universe had a specific beginning, they have been wondering: "How will our universe end?"

There seem to be three possibilities. The universe could expand forever until all the light is gone from all the galaxies and every star is a cinder; the expansion of the universe could come to a halt and reverse itself, with all the matter of the universe imploding in on itself in a horrific implosion; or somehow the expansion of the universe could reach a delicate balance at which it slows down but never quite reverses.

In the last few decades physicists have learned that the expansion of the universe is not slowing down but rather is accelerating. Something unknown is pushing the universe farther apart. Scientists are calling this unknown antigravitational force "dark energy," or the energy of nothingness. They also believe there is something called "dark matter," unlike anything on Earth. No one knows yet what dark matter and energy are; scientists currently think that they may constitute more than 90 percent of the universe. The search has only just begun.

2

Living Earth

(4.6 Billion–5 Million Years Ago)

L ife on Earth is a shimmering mystery. How can we say when life began, when in some sense our whole planet has been alive since its inception? As described in chapter 1, Earth has always managed to preserve a balance of its matter and its energy, not solidifying into coldness nor vaporizing into hotness. Helped, perhaps determined, by its size and its distance from the sun, Earth has maintained an ongoing but always changing composition.

Scientists call this active self-maintenance *autopoiesis* (Greek for "self-making"). This is the most basic definition of life—that a living organism must be able to maintain its stability while undergoing change. According to Gaia theory, Earth itself is alive because it maintains its essential stability through constant change and development. (Gaia was the Greek goddess of the Earth.) Even if Earth as a whole is not self-regulating, at least the atmosphere and surface sediments seem to form a self-regulating system, maintaining the composition of the air and the temperature of the surface for continuing life.[1]

In this continuity of development, when did Earth begin to harbor living organisms that could reproduce themselves? For years, scientists traced the beginning of life to trilobites, extinct invertebrate animals, because their fossils were the oldest records that could be found of living things. They were the first animals with hard parts, which left clear imprints in the limestone at the bottom of the seas. Their fossils, dating to some 580 million years ago, have been found in abundance all over the world.

In 1943, however, the invention of the electron microscope made it possible to see the cells that fossilize. Now it has been documented that life began as bacteria cells within the first three-quarters of a billion years after Earth formed.

Even without using an electron microscope, however, we can imagine the grand sweep of the history of life on Earth through our own bodies, a museum of that history. We are matter and energy, like the universe. Our cells, composed of atoms made in explosions of stars, maintain a hydrogen- and carbon-rich environment, like that of Earth when life began. Carbon combines with five other elements to form the chemical common denominators of all life, accounting for 99 percent of the dry weight of all living things, including us.

Every human life begins as a single cell, replaying the fact that all life on Earth began as a single cell. The first cells were bacteria, and our bodies contain ten times more bacteria cells than animal cells. Our cells contain three structures (mitochondria, plastids, and undulipodia) that evolved as separate bacteria before they were incorporated into our more complex cells.

Our blood still has the salt content of seawater; we cry and sweat seawater, testimony to the fact that all life began in the seas. Our children grow and develop for nine months in a watery environment; no life on Earth can develop its initial stages except in a wet place. As embryos, our babies still develop temporary gills—which look like tiny scars behind an embryo's ears—as a step toward developing lungs for breathing. Our bodies, like the surface of the planet, are 65 percent water. We belong to Earth in the deepest and most fundamental ways.[2]

Cells and Life Processes (3.9–2 Billion Years Ago)

How did the chemicals of the early Earth come alive? Scientists cannot say for certain, since they have not yet been able to create life out of chemicals in a laboratory. They have been trying for only about fifty years, and the planet took at least half a billion years to do it. Scientists know enough, however, to lay out the following scenario with high confidence that the missing pieces, when found, will confirm its general outline.[3]

Earth formed about 4.6 billion years ago. For the next half billion years, Earth remained a molten lava fireball, so hot that there was no surface and no water, since water could not condense but only remain a vapor high in the atmosphere (Fig. 2.1).

During these first half billion years, Earth cooled; by about 3.9 billion years ago it had cooled enough for a thin crust of rocks to form on the still-

2.1   The Surface of the Earth About 4 Billion Years Ago
(*Source:* Lynn Margulis and Dorion Sagan, *Microcosmos: Four Billion Years of Evolution from our Microbial Ancestors,* © 1997 The Regents of the University of California, Berkeley, CA: University of California Press, 39.)

molten mantle. The oldest datable rocks come from Greenland, dating about 3.8 billion years old. Volcanoes erupted at cracks and spewed lava. Meteorites crashed. Electrical storms raged. Water began to condense; it rained torrentially for perhaps millions of years. The movement of the plates of rock crust released gases from the Earth's interior, creating a new atmosphere of water vapor, nitrogen, argon, neon, and carbon dioxide. This event is sometimes called the big belch!

Somehow living organisms arose out of these conditions, within about the first 800 million years, since the oldest bacterial fossils date to 3.5 billion years old. Scientists used to think that when lightning struck the oceans, it somehow triggered living cells out of the primordial chemical soup. Now it seems unlikely that small molecules in the soup could assemble themselves spontaneously, even with the help of lightning. Scientists have imagined various explanations of how molecules could assemble themselves; the most persuasive one posits that before there were cells, there were proto-cells, or bubbles.

These bubbles formed when certain molecules assembled into primitive membranes that enclosed a tiny area; there the evolution of chemical complexity could take place. The membranes let in certain molecules and kept others out. When a bubble grew too large to maintain itself, it broke up into smaller bubbles. Bubbles differed in their molecular content. Different ones would collide and fuse. The ones that worked chemically continued; the others vanished. These proto-cells began about 3.9 billion years ago and seem to have been the vehicle for the development of molecular and metabolic complexity—a continuum of the unfolding of life.

At the earliest stage the elements in the bubbles were carbon (C), hydrogen (H), oxygen (O), phosphorous (P), and possibly sulfur (S). When nitrogen (N) entered the system, possibly as ammonia (NH_3), a dramatic increase in complexity became possible, because nitrogen is necessary for two features of cellular life—catalysis and information storage. This probably happened about 3.8 billion years ago, after 100 million years of bubbles bouncing around—the time of the emergence of the universal ancestor, either a single cell or clump of cells, from which all subsequent life on Earth descended. The evidence for a single common ancestor is the fact that all life-forms share the same genetic code, the same biochemical network. This event is sometimes called the big birth.

The universal ancestor cell took the final step from bubbles to true living cells by developing proteins, nucleic acids, and the genetic code. These first living cells were enclosed in a membrane, had about 5,000 proteins, and had strands of both ribonucleic acid (RNA) and deoxyribonucleic acid (DNA) floating around inside. The cells were about a millionth of a meter in diameter and both maintained and reproduced themselves, using their RNA and DNA to replicate RNA and DNA molecules and to cause the assembly of proteins.

The RNA may have developed first, because it can replicate itself and also act as an enzyme, evolving into cells later. The details of how this final step to life happened are still mysterious, the chemical networks are so complex that new mathematical concepts are needed before they can be understood. This ancestral cell was either a heat-loving bacterium (archaea) living off energy in volcanic vents, or a bacterial cell closely related to contemporary blue-green bacteria. The oldest fossils now existing are rocks dated at 3.4 billion years old from a mountain in South Africa that show thin, microscopic filaments similar to today's blue-green bacteria.

What does it mean to say that we evolved from archaea or from blue-green bacteria? What does "evolved" mean, more precisely? Ever since

Charles Darwin presented his theories of evolution in 1859, scientists have been discussing this question. Since bacteria evolve in more complicated ways than do more complex organisms, new understandings from the study of bacteria now suggest several ways that organisms change and develop over time.

The first, formulated by Darwin, is that random mutations, or changes, occur from one generation to the next; we know now that mutations occur spontaneously and during gene replication. (A gene is a segment of the DNA that programs a complete protein or part of a protein.) A mutation is simply a change in the sequence of nucleotides (small molecular components of nucleic acids) in a genome (an entire set of genes), which changes the instructions for making an organism. The resulting new organism passes on its genes with greater frequency than other similar organisms only if the mutation offers it some advantage in competing for resources in the environment and in reproducing itself. What Darwin figured out is the mechanism of evolution, which is adaptation to the changing environment by means of random gene mutation.

A second way to evolve is the way of bacteria. Bacteria reproduce by growing to twice their size, replicating their simple strand of DNA, and dividing, with a strand of DNA in each new cell. Fast bacteria divide every twenty minutes or so. If threatened, bacteria spill their genetic material into their environment, and other bacteria pick up some of the pieces. Bacteria recombine their DNA, as humans are just learning to do. Bacteria may change 15 percent of their genetic material every day. They form a planetary web that can exchange genetic material extremely rapidly.

A third way of evolution is called symbiogenesis. This happens when a symbiotic arrangement of two organisms becomes a permanent one. One example close to home is the bacteria that cannot function in oxygen but still live inside our intestines, where there is no oxygen, to assist in the absorption of our food.

For 2 billion years, from 3.8 to 1.8 billion years ago, bacteria worked their mysterious ways. During this immense period of time, bacteria created fermentation, nitrogen fixation, photosynthesis, movement, and the fundamentals of the planetary ecosystem.

In the beginning, when the first living cells did not have enough genes to take care of all the amino acids, nucleotides, vitamins, and enzymes that they needed, they absorbed components directly from the environment. As bacteria increased and began using up nutrients, the ones that survived had to develop new metabolic ways to extract food and energy from materials at hand. In one of the first innovations, bacteria began to convert sugar to energy. Other bacteria, living in mud and water away from sun-light, developed ways of breaking down sugar (fermentation) that are still in use today. Some bacteria evolved the ability to take nitrogen gas from the atmosphere and convert it to a chain of amino acids. All organisms today depend on a small group of bacteria that are able to fix nitrogen from the air.

Bacteria also evolved the process of photosynthesis, or converting sun-light and carbon dioxide from the air into food. Early photosynthetic bac-teria took hydrogen directly from the atmosphere to combine with carbon to form carbohydrates. This metabolic innovation by bacteria, not yet fully understood, ranks as one of the most important in the history of life on the planet. Bacteria were also able to cycle gases and soluble compounds through the Earth's atmosphere and water to exert some regulation of the conditions in which they lived.

As bacteria approached 2 billion years of life (1.8 billion years ago), they covered every available nook of the Earth's surface. Shallow pools abounded, brilliantly colored purple and sienna by floating bacteria. Greenish and brownish patches of scum floated on waters, stuck to banks, and tinted damp soil. Volcanoes still smoked in the background, and the air was filled with stinky odors emitted by layers of bacteria piled up to create living rugs. By then bacteria had probably evolved all the major metabolic and enzymatic systems, but they were still cells without a nu-cleus, or prokaryotes. Their genes floated freely inside; they were not yet packed into chromosomes wrapped in a membrane to form a nucleus. Even so, bacteria had already established the fundamentals of the plane-tary system.

New Cells and Two-Partner Sex (1.8 Billion–460 Million Years Ago)

About 2 billion years ago Earth underwent a catastrophic pollution crisis. Earlier there had been almost no oxygen in the air, but gradually so much oxygen was released into the atmosphere by blue-green bacteria taking the hydrogen from water that the oxygen, which they could not use, threatened all bacteria. Oxygen was toxic because it reacts with the basic compounds of life (carbon, hydrogen, sulfur, and nitrogen). Oxygen in the atmosphere rose gradually from one part in a million to one part in five, or from 0.0001 to 21 percent.

The bacteria that survived this change in the atmosphere had to reor-ganize on a massive scale. In one of the great coups of all time, blue-green bacteria invented a way to breathe oxygen, using it in a controlled manner. They now carried out both photosynthesis, which generated oxygen, and respiration, which consumed it. The level of oxygen in the atmosphere stabilized at about 21 percent, which is the present level. How that level is maintained is still a mystery, but if the level of oxygen were a few percentage points higher, living organisms would combust; if it were a bit lower, organisms would asphyxiate.

As the level of oxygen in the atmosphere was rising to 21 percent, a new kind of cell arose. The bacteria that evolved a way to breathe oxygen had tapped into an energy source that was far beyond their ability to fully use. Some of them evolved a new kind of cell, called a eukaryote ("truly nucleated"), which has two defining features: a nucleus wrapped in its own membrane and an oxygen-using part called the mitochondria. Many see this leap from cells with no nucleus to cells with a nucleus as the most dramatic move in all biology. It has never happened again; all the multicellular creatures of today are composites of cells with nuclei (Fig. 2.2).

The new cells, much larger and more complex than cells with no nucleus, had pulsing cytoplasm streaming around their internal structures. Inside their nuclei were chromosomes containing 1,000 times more DNA than that in no-nucleus cells. Just what the function is of this enormous amount of DNA is one of the puzzles of molecular biology. Some of the new cells also had photosynthetic parts called plastids, or chloroplasts, along with the oxygen-using parts called the mitochondria. It seems likely to many biologists that these two parts represent once-separate bacteria that got trapped inside other bacteria. A lab experiment has shown that in amoebas—microscopic one-celled animals—dangerous bacteria can become required elements (organelles) in less than a decade. In a like manner, the cell with a nucleus seems to be a merger of diverse organisms.[4]

Eukaryotes appeared at the earliest 1.9 billion years ago. Between 1.7 and 1.5 billion years ago, the organisms formed by these cells with nuclei evolved a new way of reproducing, which required two partners. The sperm cell from one organism joined the egg cell from another. After they combined and divided, a new organism emerged with a complete set of chromosomes, half from each parent. This kind of sexual reproduction has continued unchanged to its current manifestations.

With the new nucleated cells and the new sexual reproduction, cells began to clump together more frequently. The old cells had sometimes stuck together as multicellular organisms, but the new cells began to stick together more spectacularly, and eventually became plants and animals. Two sex cells combining to form a new one produced more genetic

PROKARYOTIC CELL   **EUKARYOTIC CELL**

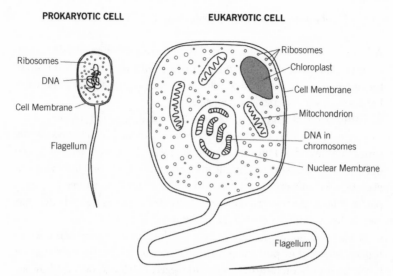

2.2   Prokaryotic and Eukaryotic Cells Compared

Cells are biochemical factories surrounded by porous membrane and include genetic material (DNA) that codes cellular functions and reproduction. Ribosomes are structures where proteins are assembled, following the instructions from the DNA. The more complex eukaryotic cell has a genome comprised of strands of DNA within a membrane that forms a nucleus. The area outside the nucleus is crisscrossed by more membrane, which organizes the different organelles in the cell. One kind of organelle is the mitochondria, which converts food into chemical energy; another kind is the plastids, or chloroplasts, which convert light into chemical energy, known as photosynthesis. Eukaryotic cells have a whiplike flagellum for movement.

variability—from the recombination of genes from two sources and from the mistakes in copying (mutations), both of which increased the possibilities for new organisms.

Five-sixths of the history of life is the story of one-celled creatures, bacteria. They created all the chemical structures that make our life possible. We often see bacteria as germs to be conquered, but they are also ancestors to be honored—not to mention guests to be entertained; each of us has about a trillion of them grazing on our skin. They still rule, as they always have; the smaller a living entity, the easier it is to form and maintain because it is less complex.

Plants and the Face of the Earth (460–250 Million Years Ago)

As we have seen, living cells survive in a close relationship with their environment. Indeed, biologists have trouble agreeing on a clear definition of
the difference between life and nonlife because of the vital bond between
Earth's environment and its organisms.[5]

After single blue-green bacteria cells came alive, groups of them stuck
together and became colonies. They lived in wet, sunlit shallows. Sometimes the water dried up, and some colonies of blue-green bacteria developed the ability to stay wet on the inside and dry on the outside. With this
evolutionary advantage, they survived and multiplied to become early
plants, related to modern-day mosses and liverworts. By 460 million years
ago the first plant spores had come ashore.

Once on land, plants had to become three-dimensional and develop
tough stems to transport water up from roots and food down from the flat-
tened ends of branches that were early leaves. This had occurred by 400
million years ago. Next came seeds, to protect plant embryos from drying
out when water was not available. Seeds enabled plant embryos to pause,
monitor the environment, and wait for favorable conditions before continuing their development. Next arose fern "trees," which, from 345 to 225 million years ago, covered the land areas of Earth.

Exactly what shape did the land areas of Earth take? For a long time
people assumed that the continents as we know them today had always
been stable, in the same place. Now we know otherwise.[6]

The Earth is a huge electromagnetic dynamo. Its inner core is solid in
the center, surrounded by liquid iron and nickel, still heated from the primordial creation of the planet. The magnetic field of Earth is generated by
the rotating liquid iron swirling around its center. The solid core slowly increases in diameter two inches every five years because, like everything in
the universe, Earth is cooling.

A layer of partially melted rock, called magma, reaches up from a depth
of a hundred miles to the crust of solid rock on the Earth's surface, which is
four to twenty miles in thickness. This crust of rock covers the whole of
Earth; continents are upward bulges in the crust and oceans are shallow
dents of about two miles depth filled with water. The crust cracks into
tectonic plates which push into, over, and under each other as they float on
the magma. Heat in the magma forces it up through midocean centers,
through volcanoes onto the surface, and into cracks between the plates of
the crust, causing earthquakes as it forms new crust and moves the exist-

ing crust around. The rocky surface of Earth has eroded and been deposited on ocean floors, consolidated into rock, and thrust up again about
twenty-five times in Earth's history.[7]

The continents of the Earth are carried along on its moving magma at
speeds measured in inches per year. The movement of continents through
time can be studied, because when new rocks are created, their magnetism
aligns with that of the north and south magnetic poles as they were when
the rock was formed. (The study of this phenomenon is called paleomag-
netism.) Since the magnetic poles move a little every year, scientists can
determine how much the rocks have moved and when the rocks were laid
down.

By combining data about the movement of tectonic plates with the pa-
leomagnetic and fossil record, geographers have reconstructed the position of the Earth's crust over time. Of course, the farther back in time they
go, the less certain they are and the more arguments and interpretation
abound. Yet all agree that Earth offers no long-range, permanent conditions to any of its life-forms.

It seems that by about 250 million years ago, when seed ferns flourished, most of our current continents had moved together, huddling toward the South Pole in one massive continent called Pangaea (All World).
Earlier than this, landmasses had floated about in fragmented islands, with
much of present land under water; even earlier the landmasses may have
been more nearly unified.

Pangaea lasted some 50 million years before splitting up again into a
top half called Laurasia (North America, Europe, and Siberia) and a bottom
half called Gondwana (southern hemisphere). Gondwana later split into
South America, Africa, Madagascar, Arabia, Australia, Antarctica, and
India. The continental plates will keep moving as long as the Earth's core is
hot from the heat generated during its formation and sustained by decay-
ing radioactive elements (Fig. 2.3).

After Pangaea began to split up, seed fern trees developed into conifers
and then into flowering plants and hardwood trees, which appeared about
100 million years ago. The earliest modern families to appear were beech,
birch, fig, holly, oak, sycamore, magnolia, palm, walnut, and willows. Red-
wood trees watched over the dinosaurs.

Trees, and other plants, played (and still play) a major role in keeping
Earth cool for other life. Every day Earth receives huge torrents of energy
from the sun—the equivalent of 100 million Hiroshima bombs' worth. It
also receives fresh infusions daily from its internal core. Most of the energy

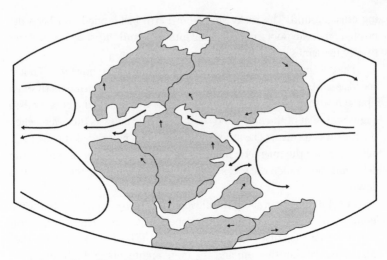

2.3   The Breakup of Pangaea About 200 Million Years Ago

that Earth receives from the sun bounces off and is reflected back into space. Plants convert a small percentage of solar energy through photosynthesis, but they help most by removing carbon dioxide from the air. This is cooling because, while atmospheric carbon dioxide is permeable to the incoming energy from the Sun, it resists letting heat escape back into space. The atmosphere is only about .035 percent carbon dioxide, but this tiny percentage is critical to stabilizing the temperature on Earth. The photosynthesis of plants also releases oxygen into the atmosphere, helping to maintain the percentage at about 21 percent, critical for living organisms.

Animals Come Ashore (450–65 Million Years Ago)

In summary, by about 250 million years ago bacteria on shore had dried out and developed into giant seed ferns that covered one huge continent, Pangaea. What about animals? When did they get four feet on the ground?

Animals began in the sea. Even before plant cells began to clump together in shallow water, animals began to develop in the ocean waters. Animals are distinguished from plants by the specialization of cell function and the complex interaction among cells. Animal cells are bound to neighboring cells by a great variety of elegant intercell connections, only recently visible via electron microscopes. These esoteric cell junctions are now considered the true mark of animality, along with the ball of cells that becomes an embryo. Also, animal cells have no photosynthesis organelles incorporated in them. Animals took the complexity of interconnections among cells and ran with it to new levels of extravagance.

The line to animals first began when one cell, with a nucleus and with cell whips for locomotion but without photosynthesis, first stuck to another cell and propelled it along, so that the second cell could use its microtubules for some other function. The simplest animal that now exists, *Trichoplax*. It is a little colony of nucleated cells, 3 mm across, walking on cell whips, nothing more.[8]

Since animals first developed as soft-bodied creatures in the sea, what evidence could there be of any early animals? In 2004 tiny fossils of bilateral flat bodies, the width of four human hairs, were found in rock in southwest China, dated at about 600 million years ago.

By 580 million years ago animals had developed hard parts—shells and external skeletons visible to the naked eye—and their fossils could be found in abundance all over the world. By this time our bacteria ancestors had already been around for about 3 billion years. The earliest animals with hard parts were trilobites and giant sea scorpions, sometimes reaching more than three meters in length. All of these are now completely gone, extinct along with probably 99 percent of all the species that have ever lived.[9]

Animals took slightly longer to go ashore than plants, probably because of the size and complexity they developed in the sea. The coming ashore of animals is thought to have begun about 460 million years ago, with a creature something like a sow bug likely to have been the first. Why did they dare? Probably because it was getting dangerous in the water. Sharks had evolved, and Pangaea was forming, resulting in much less coastline for easy feeding. While amphibians need to be submerged in water for some part of their life cycle, reptiles, birds, and most mammals, except as embryos, do not.

Another life-form, the fungi, also came ashore. Neither plant nor animal, fungi represent a third fundamental way that nucleated cells evolved. Fungi develop from spores, their cells may contain many nuclei per cell, and they get nutrients by absorbing molecules directly from soil or wood rather than by eating or by photosynthesis. The part of the fungus we see is the "fruit"; the body is actually a web of tiny filaments underground. Molds (such as penicillin), mushrooms, yeasts, morels, and truffles are common examples, most of them terrestrial. Fungi coevolved with plants and animals, all of which are deeply interconnected.

Taking the micro point of view, we see that, beneath all their wild exuberance, plants, fungi, animals, and bacteria form a shimmering landscape, a symbiotic intercommunicating network of modules of nucleated cells. Or, removing our microscope and taking the macro point of view, we see that, together, plants, fungi, animals, and bacteria form a single living community (biota) that monitors and regulates the biosphere to sustain conditions for life.

Two hundred and fifty million years ago this living community experienced threats so severe that in a period of less than several hundred thousand years it lost more than 50 percent of its families and up to 95 percent of its species. The biota succeeded in sustaining life, but only with this extreme loss. (In the classification system, the category with the most entries is that of species who can breed together. Species are then grouped into genera, genera into families, families into orders, and so on up.)

What happened to wipe out more than 50 percent of the families on Earth? Mass extinctions, we now realize, have occurred repeatedly in the history of the planet, at least five or six times. How often, and whether with any regularity, is still debated. But there is agreement that the extinction of 250 million years ago was the harshest extinction that has ever occurred.

Scientists have collected much data and generated many theories about the extinction of 250 million years ago, but definite conclusions are still premature. The most likely suspects for its cause are changes in sea level, atmosphere, and climate; massive volcanic eruptions; and/or extraterrestrial impacts.[10]

The continents were fusing into Pangaea in the 20 million years before the extinction; this fusing may have produced extreme climate change. Large-scale volcanic eruptions in Siberia and southern China, firmly dated at 251.1 to 252.2 million years ago, may have blocked the sun and triggered glaciation. Marine oxygen may have dropped precipitously. A huge meteor may have hit in the Indian Ocean northwest of Australia. The debate continues.[11]

After a mass extinction, life seems to respond by creating many new forms of life much more rapidly than usual. These new forms fill the niches left by the extinct forms. Before the extinction of 250 million years ago, amphibians were the dominant animal life; some had already developed into reptiles. After the extinction, reptiles multiplied and developed rapidly into astonishing new species.

Amphibians became reptiles by developing a closed egg, which could be laid on land with no need for the parents to return to water. To accomplish this, reptiles had to develop penetrative sex, in which the male could deposit sperm inside the female instead of fertilizing the eggs after they were laid. We have much for which to be grateful to reptiles.

Within 25 million years after the extinction, reptiles had developed into the amazing creatures we call dinosaurs. They rose to worldwide dominance around 210 million years ago, before Pangaea began breaking up about 200 million years ago. For well over 100 million years every other creature lived in the byways of the dinosaurs. Geologists call the eras in which they lived Cretaceous, Jurassic, and Triassic.

There's a lot of guesswork about dinosaurs, but experts agree that they were a single group from a single common ancestor, that they were mostly terrestrial, and that birds are direct descendants from one group of meat-eating dinosaurs. Dinosaurs ranged in size from about two feet long and weighing only five pounds, up to giants like the *Brachiosaurus* that stood thirty-five feet high and weighed as much as seventy tons. Dinosaurs had the run of the world while Pangaea lasted, but the landmass began to split apart during the creatures' heyday. The most famous kind of dinosaur, *Tyrannosaurus rex*, made its appearance late in the age of the dinosaurs. It was the largest meat-eating land animal ever, at forty-six feet in length, twenty feet in height, five tons of weight, sporting teeth six inches long, probably a scavenger rather than a predator. A seven-year-old child, if one had been around, could have stood inside the gaping mouth of a *Tyrannosaurus rex*.

People today are fascinated by the world of dinosaurs because of their exciting size, variety, and dominance, and because their world begins to be familiar to us. Their waters were filled with fish and amphibians. Their climate was tropical, with lush plants, flowers, and bees, so attractive to us today. Dinosaurs developed mating patterns, and some were even beginning to care for their eggs and offspring. Little furry mammals, much like our childhood stuffed animals and pets, were running underfoot, coming out at night to hunt for plants and animals. Only a few elements were missing from the familiar picture. Birds were just beginning to develop from winged dinosaurs, and there were no cave people yet, not for another 62 million years after the end of the dinosaurs—not even any great apes for another 35 million years.

Dinosaurs to Chimpanzees (65–5 Million Years Ago)

Right in the middle of the dinosaurs' prime, when they ruled the world with an incredible range of diverse types, another extinction occurred. This one,

at 65 million years ago, wiped out all the dinosaurs (except those developing into birds) and every other land animal that weighed more than fifty-
five pounds. This extinction proved no worse than some previous ones, but
it stays in our imagination more vividly, perhaps because we identify more
easily with the deaths of dinosaurs than we do with the deaths of earth-
worms, trilobites, or microorganisms.

Some of the groups that became extinct 65 million years ago seem to
have vanished suddenly. Other groups showed gradual reductions in their
diversity from 75 to 65 million years ago. The survivors of this extinction
seem to have been most of the land plants and the small land animals—
insects, snails, frogs, salamanders, turtles, lizards, snakes, crocodiles, some
placental mammals, most of the fishes, and marine invertebrates.

Scientists have made some wild guesses in the past about the causes of
the dinosaurs' extinction—that the dinosaurs were too stupid, or too con-
stipated, or that small furry mammals stole their eggs. These ideas have
been rejected in the face of evidence of an impact catastrophe, in which a
six-mile-wide meteorite struck Earth, kicking up so much debris that the
sun's rays may have been blocked for thousands of years. In 1991 geologists
identified the crater, 120 miles wide and twenty miles deep, which now lies
buried beneath the Yucatan Peninsula in Mexico. Named for the village located near it, Chicxulub (pronounced cheek-shoe-lube), the crater was
formed on the north coast of the Yucatan at the water's edge, where it sent
tsunami waves racing across the Gulf of Mexico. Many volcanic eruptions
occurred at the same time, as also happened at the time of the extinction
250 million years ago, but geologists do not yet understand the connections between volcanism, extraterrestrial impacts, and extinctions.[12]

As mentioned earlier, paleontologists used to think that small mammals might have caused the extinction of dinosaurs by eating their eggs;
now they think that the rise of mammals was an effect of the demise of the
dinosaurs, which left the planet wide open for mammals to develop. There
was so much ecospace to be filled that mammals were able to develop an
extreme diversity of types.

Mammals are defined by their ability to deliver their young alive, either
as tiny ones who crawl into an external pouch for further development (the
marsupials), or as larger ones who develop internally (placentals). The earliest mammals appeared about 210 million years ago and, until the extinction of 65 million years ago, few were larger than rats. They had fur, which
kept them warm, and ate insects and flesh; later some added plants to their
diet.

Now we know, however, that the most important distinguishing feature of mammals was the development, between 150 million and 100 million years ago, of a new area of its brain, the limbic area. The limbic area of the brain monitors the external world and the internal bodily environment and orchestrates their congruence. It fine-tunes physiology to adjust the body to the outside world, thus enabling mammals to stay warm in cold places. It is the seat of emotions and orchestrates the facial muscles that express emotions.

The presence of the large dinosaurs constituted good luck for the evolution of the mammals. It kept most of them small and close to the ground, where they developed better teeth, smell, and hearing as night creatures that foraged while the dinosaurs slept. Squirrels and shrews are the best current representatives of the mammals that scampered beneath dinosaurs.

After the dinosaurs expired, it took mammals several million years to develop even moderately large bodies. The history of living creatures becomes fragmented during this time because the supercontinent of Pangaea was breaking up into smaller pieces. The isolated continents meant that animals could not travel across one landmass; diverse forms developed on each continent. Most of the early large mammals would seem lumbering and awkward to our eyes; they grew in heavy woodlands, not the open grasslands that later produced the graceful, long-legged forms adapted to running.

Earth's climate kept changing, providing the impetus for new evolutionary developments. From 55 to 50 million years ago the temperature grew warmer; there were jungles at the poles. Two large mammals, whales and dolphins, returned to the sea.

By 35 million years ago the annual temperatures began to plunge, as more land area split up—Australia from Antarctica and Greenland from Norway—redirecting the flow of ocean currents. The resulting clash of cold water with warm water produced cooler weather. Many animal groups were lost; new ones arose. Early primates—little lemurs, bush babies, and monkeys—survived in tropical areas, which had a year-round supply of fruit. During the first 5 million years of this cool era, the first apes appeared.

After another 10 to 12 million years (by 23 million years ago), the temperature began shifting again to warmth. Tectonic plate pressures produced the Cordillera mountains of North America (the Rocky Mountains, the Coast Range, the Sierra Nevada, the Sierra Madre) and the Andes of South America. The whole continent of India rammed into Eurasia to produce the Himalayas. The continent of Africa became connected to Eurasia, allowing the unique African animals to enter, especially elephantlike creatures and apes.

By 10 million years ago the temperature became the warmest it had been since 35 million years ago. Then it shifted to cool again, with a loss of carbon dioxide in the atmosphere and a reverse greenhouse effect. As a result of these changes, grasslands appeared in North and South America, considered a key event in the last 500 million years. Grass covered one-third of the world's surface and became the staple food of the animal population. Animals who could digest grass were guaranteed a self-replacing supply of food. The development of grasslands and its collectivity of animals in the Americas 10 to 8 million years ago foreshadowed what was to come in the savannah grasslands of east Africa in the last 7 to 5 million years.

By now it should be clear that the shifting climate underlies Earth's story. The climate changes seem to have been generated mostly by the continents moving about on Earth's mantle of molten magma, creating mountains and changing the flow of ocean waters. Striking meteorites probably also affected the climate, as well as changes in the Earth's tilt, wobble, and orbit—a complex interacting network of many factors.

Small primates—or mammals with flexible hands and feet with five digits, nails, and forward-facing eyes—first appeared some 55 to 60 million years ago. By about 25 million years ago, some had developed into larger animals, called hominoid primates, or apes. Apes developed for 20 to 25 million years until the split between the human group and the ape group occurred, probably between 5 and 7 million years ago, much more recently than previously believed (Fig. 2.4).

The only evidence available for verifying the story of ape/human development consists of extremely fragile fossilized bones and footprints that are millions of years old and scattered from place to place, without complete records anywhere. There is no way yet to present a complete family tree—too many gaps remain in the evidence, even though it has improved markedly in the past twenty years. There are two large gaps in the fossil record: from 31 to 22 million years ago, when gorillas, chimpanzees, and humans launched their careers, and from 12 to 5 million years ago, when the great apes and humans diverged.

The first primates developed in the tropics and subtropics, where they mostly were tree-dwelling monkeys. Basic to their characteristics are the five digits on each of four limbs, with nails instead of claws, and opposable thumbs and usually opposable toes. Monkeys also had eyes that faced for-

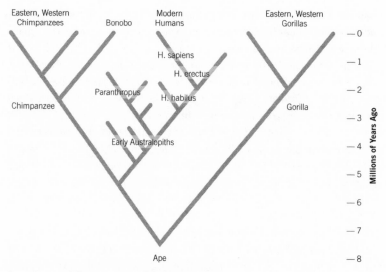

2.4   Apes and Humans: A Simplified Family Tree

ward rather than to the sides, for more overlapping fields of vision. Since their brains had to coordinate the overlapping fields to produce depth perception, they developed larger brains than other mammals. They gave birth to one offspring at a time, after a long carrying time; their babies' slow growth and dependency resulted in complex social arrangements in order to support infant development over long periods of time.

In contrast to monkeys in other parts of the world, the monkey primates in the Americas never came down from the trees. The reasons for this remain unknown. In Asia, Europe, and Africa some monkeys ventured down to become apes, or hominoid primates, that developed into humans. They showed up in Africa about 25 million years ago, and appeared by 18 million years ago in southern Eurasia from France to Indonesia. The development of apes in Europe and Asia went along for millions of years but eventually ran into difficulties. In Asia only one great ape, the orangutan, survived. In Europe the climate became drier and eliminated early hominoids by about 8 million years ago. Only in eastern Africa did the great apes continue to thrive and innovate.

What is so special about eastern Africa? It features the Rift Valley System, a fracture in the African continental plate that runs 4,000 miles (6,436 km) from Ethiopia and the Red Sea in the north down through Kenya, Uganda, Tanzania, and Malawi into Mozambique in the south. For 20 mil-

lion years tectonic activity along this break in the plate produced volcanoes, raised highlands, and collapsed lowlands to form valleys that chan-
neled water into the continent's largest lakes. Every kind of climate is
present—tropical forests give way to open woodlands, which open to savanna grasslands. There are altered patterns of rainfall and geographic bar-
riers, which can isolate groups of animals. The Rift Valley System provided
the perfect lab for evolutionary experimentation.

The African great apes consisted of two species of chimpanzees (the
common chimpanzee and the bonobos, formerly called pygmy chimps),
and two subspecies of gorillas. Recent studies have concluded that humans
share 98.4 percent of our DNA with chimpanzees, who constitute our next
of kin. (For comparison, we share about 90 percent of our genes with the
rest of the living world.)[13]

The great apes have been studied only since the early 1960s, when Jane
Goodall went to Tanzania to observe chimpanzees in the wild. Up to that
time no one paid much attention to chimpanzees, except in zoos, nor knew
much about them. After Goodall began to teach us about chimps, people
started to realize that only through these animals could early human history be understood. If chimpanzees, bonobos, and gorillas had become extinct before scientists had begun to understand evolution, it would now be
impossible to imagine what early humans were like.[14]

After forty-five years of diligent study, there is an emerging consensus
among biologists about chimpanzee behavior. The two species (common
chimps and bonobos) vary strikingly in their behavior. Common chimps
live in territories whose borders are aggressively defended by males. Males,
when they grow up, stay in the same territory, while females move to another. Males and females live in separate hierarchies, not in conjugal pairs.
Males make females defer to them, with violence when necessary. Both
males and females have multiple mates. Chimps have the verbal intelli-
gence of a small human child, and each has a distinct personality and tal-
ents. They can learn sign language and use it to converse with each other
and with people, and they can teach it to their children. The basic diet of
apes is fruits and plants, but chimpanzees also love raw meat and kill vi-
ciously for it. Chimp mothers form durable bonds with their children, while
males show little interest in child rearing. Chimps are highly social, living in
groups of eighty to a hundred. Their emotional life is closely related to that
of humans; they get angry, jealous, anxious, and lonely, and are protective
of the weak and willing to share.

The other kind of chimp, the bonobo, is a different kind of creature.

Slightly smaller than a common chimpanzee, a bonobo has a proportion-ately smaller head, neck, and shoulders, with a flatter, more open face. They were not recognized as a separate species until 1929 and have been studied even more recently than chimpanzees. In the wild, bonobos live only on the south bank of the Congo River, in the Democratic Republic of the Congo (formerly Zaire). Bonobo society, much less hierarchical than the common chimp society, is controlled by females. Bonobos seldom kill; they settle conflicts by having sex in a great variety of ways. Since the two species of chimpanzees developed after the human line broke off from the chim-panzee line, we are theoretically related equally to both species.[15]

However, chimps are not humans, and vice versa. A clear genetic marker distinguishes between them: chimps have a total of forty-eight chromosomes (twenty-four pairs), while humans have forty-six (twenty-three pairs). Chimps are quite unlike humans in many other distinctive ways. They take only ten to fifteen seconds to copulate, cannot distin-guish between legal and illegal behavior, cannot speak, and, when they learn from humans to sign, they "converse" only at the level of a two-year-old human.

Remaining in the Great Rift Valley in eastern Africa, we turn in the next chapter to how humans evolved in the 5 to 7 million years since people and chimpanzees diverged from their common ape ancestor.

Unanswered Questions

Much of our knowledge of the span of time that this chapter covers—almost 4 billion years—is based on unavoidably incomplete evidence. Many questions remain unanswered, but new evidence, as it comes in, is confirming the fundamental story.

Recent advances in dating rocks and fossils have helped. They rely on radioactivity, or on the statistical tendency of the nucleus in some versions of an element, called an isotope, to change in random ways. The nucleus of an isotope is unstable, or radioactive. The time it takes for one-half of an amount of an isotope to decay, or change its nucleus, is called its half-life. Volcanic rocks often contain radioactive isotopes, which provide a way to date the rock. Called radiocarbon dating, this process was invented in 1948 and recently improved.[16]

1. Were dinosaurs warm-blooded?

Robert Bakker, in his book, *The Dinosaur Heresies*, explains why he thinks dinosaurs must have been warm-blooded in order to dominate for so long. Sharing warm blood with mammals, a leap from other reptiles, would have given the dinosaurs their competitive edge. But this powerful idea cannot be proved or disproved, since fossils do not contain any evidence of dinosaurs' organs or how they worked.

2. How should living things be classified?

With increased knowledge of microorganisms, scientists are proposing whole new systems of classification, giving more room to divisions among bacteria. In 1969 an ecologist from Cornell University, R. H. Whittaker, proposed five main kingdoms—Animalia, Plantae, Fungi, Monera (bacteria), and Protista (anything not plant or animal). By 1976 Carl Woese proposed twenty-three main divisions, grouped under a new level called domains—Bacteria, Archaea, and Eukarya. On this model all of botany and zoology "are relegated to a few twigs on the outermost branch of the Eukaryan limb."[17] Thus the battle for classification proceeds.

3. How does evolution help explain human nature?

Evolutionary psychologists treat the mind as a feature of humanity adapted to survival. Their approach has been a trend in thinking in the last few decades. They argue that our brains have evolved from chimpanzee brains and from the situation of at least the last 2 million years ago, rather than from our experience of the last 5,000 years, which has not yet been encoded in our genetic material. One small example is our fear of snakes and spiders, which they argue is innate and was put in place during our development in snake- and spider-ridden places. Apes who were afraid of snakes and spiders tended to survive, and that is built into our brains by the deaths of those who were unafraid.[18]

4. If chimpanzees are such close relatives to humans, the question arises: Has a human ever tried to mate with a chimp? If anyone has tried this, they have kept quiet about it. No such experiment is known. Chimps and people are different species, and they should not be able to produce offspring. If any such experiment succeeded, how would the baby be raised?

The fate of the great apes is hanging in the balance. They have lost nearly all their forest world, are dying from Ebola virus, and people hunt them for meat, keep them in zoos, and put them in cages for medical research. Those who have studied them most closely are heartbroken. The African people who live nearest them often find their interests in conflict with those of the apes. In the lifetime of today's children, apes are likely to cease to exist in the wild.

5. Some biologists and astronomers are determined to find other life in

the universe—or to spread life from Earth. They ask questions like: Will the microcosmos—the world of microorganisms, or bacteria—ever spread to other places in space? Could colonies of bacteria create the conditions for life elsewhere? Will bacteria be taken to other planets to start colonies there?

3

Human Emergence: One Species

(5 Million–35,000 Years Ago)

It is time to get a handle on the eons that have elapsed in our story. Those who come from a Judeo-Christian heritage may have learned to think of the world as being created only several thousand years ago—beginning in 3761 BCE according to Jewish reckoning or in 4004 BCE according to a reference note in the King James version of the Bible. Yet other cultures have a longer conception of time. Mayan inscriptions refer to a million years ago; they may refer to 40 million years ago, though this is disputed. In the Hindu religion, the cosmos itself undergoes death and rebirth; a day and a night of Brahma last 8.64 billion years, a bit more than half the time since the big bang in scientific cosmology. A Chinese astronomer, I-Hseing, in the eighth century considered the world to have been in existence for millions of years.

Since humans live only seventy to a hundred years, we have no direct experience of cosmic time. We need some analogy or metaphor to render time sufficiently concrete for us to grasp it. We must use our imagination or be limited by our brief experience of time.

To imagine time since the big bang, we can use the device of compressing all time into thirteen years. If we say the universe began thirteen years ago, Earth would have come into existence five years ago; the meteorite that killed the dinosaurs would have hit three weeks ago; the first bipedal apes (those that walked on two feet) would have appeared three days ago and the first *Homo sapiens* fifty-three minutes ago; and modern industrial societies would have existed for six seconds (Fig. 3.1).[1]

3.1   A Compressed Timeline of the Universe

If the universe had begun 13 years ago, then, at this moment . . .	
The Earth would have existed for about	5 years
Large organisms with many cells would have existed for about	7 months
The asteroids that killed off the dinosaurs would have landed	3 weeks ago
Hominids would have existed for just	3 days
Our own species, *Homo sapiens,* would have existed for	53 minutes
Agricultural societies would have existed for	5 minutes
The entire recorded history of civilization would have existed for	3 minutes
Modern industrial societies would have existed for	6 seconds

Source: David Christian, "World History in Context," *Journal of World History,* December 2003, 440.

The American Museum of Natural History in New York City recently installed an exhibit of universe history that begins with a light-show simulation of the big bang. After standing in the big bang, the visitor follows a long ramp that spirals down two floors. At the end is a plaque containing a line the width of a hair that represents 30,000 years of human history—a metaphor I cannot forget.

As I was writing this story of the world, several friends suggested that I let each page of my book represent a certain number of years. That thought occurred to them before they did the calculations, however; to represent just the 4.6 billion years of Earth history with, say, 300 pages, each page would have to represent 15 million years. And *Homo sapiens* would not enter until the last third of the last line. Most of the book would contain blank pages representing unknowable time—not a good marketing strategy.

Setting aside the earliest two-thirds of time, consider just the duration of Earth's existence. For a linear metaphor, imagine laying out a string the length of thirty-one and a quarter football fields (3,125 yards). This would represent the 4.5 billion years since Earth began. The divergence of human development from that of the great apes, at about 5 million years ago, would occur at three and a half yards from the end of the string. The leap from hominids to *Homo sapiens* would occur about five inches from the end, while the time since agriculture began would be represented by a quarter of an inch.

Another simple way to represent Earth time is to condense it to the scale of time most familiar to us—a twenty-four-hour day. If we imagine the age of the Earth as a single day beginning at midnight; then the first single-celled organism would appear at about 4 A.M. and the first sea plant

not until the evening at about 8:30 P.M. Plants and animals would get to
land at about 10 P.M., with dinosaurs appearing just before 11 P.M. Dinosaurs
would disappear at twenty-one minutes before midnight; humans would
appear less than two minutes before midnight, and agriculture and cities
just a few seconds before 12 P.M.[2]

However one represents it, the bare reality is that human history constitutes the tiniest fraction of planetary time—not to mention universe
time.

From Divergence to *Homo erectus*

There is no exact moment at which human beings began to appear; the
boundary between humans and apes is not a fixed point. Five to 7 million
years ago some mutation occurred in an ape ancestor and survived, and
from that single mutation other single mutations kept occurring in the
branch called hominids, the bipedal apes. The mutations that bestowed
advantage were preserved. These changes eventually led to modern *Homo
sapiens*.

These genetic changes took place repeatedly in the same place—
eastern Africa. For at least 3 million years human development occurred
only in Africa; hominids did not live anywhere else, although apes lived in
Europe and Asia as well. Sometime between 1 and 1.8 million years ago, a
group of hominids we call *Homo erectus* left Africa and began to spread out
over the rest of the Earth. Later, about 100,000 to 200,000 years ago, another
group of people, by then evolved into *Homo sapiens*, left eastern Africa to
inhabit the Earth, while the previous group of *Homo erectus*, as it had
evolved in various places, became extinct. This is the big picture, as best as
it can be constructed at the present time, with possibly another migration
occurring between them.[3]

Why did human evolution happen in eastern Africa? What are the
characteristics of this continent that it alone could cradle the development
of human beings?

Eastern Africa is tropical; our lack of hairiness indicates that we
evolved from tropical animals. To become humans, tropical apes came
down from trees to live on grasslands; we are creatures of grasslands,
not forests. The geography that could mold human development is
found in the Great Rift Valley of eastern Africa, as described in the previous
chapter.

People who visit the Great Rift Valley, the Olduvai Gorge, or the

Ngorongoro Crater in Tanzania are usually deeply moved by the beauty of the place and the recognition of their ancestral homeland. There on the edge of the Serengeti Plain one can still see the abundance of animal and bird life that provided sustenance as humans emerged from apes. The walls of the gorge, the wooded areas, and the open plains provided shelter, enclosure, and food for the hunter-gatherers who crouched beside them.

The Great Rift Valley is formed by a rift in the African continental plate; the eastern piece of Africa will one day split off from the rest of the continent and drift away in the Indian Ocean, eventually to crash into India, China, Japan, or who knows where. The rift begins at the Red Sea in Ethiopia and extends down through Kenya, Tanzania, and Mozambique, with branches of it extending into Zaire and Zambia. The equator runs right through the middle of the length of the rift, at Mount Kilimanjaro in Tanzania. The flat coastal plains rise to an interior plateau between 1,200 and 1,400 feet (365–1,220 meters) above sea level. These highlands maintain the temperature range that is physiologically most natural for humans, on either side of 80 degrees Fahrenheit (Fig. 3.2).

The landscape in the Rift Valley was a tropical mixture of woodlands and grasslands, or savannas, with occasional mountain ranges. During the wet months lush grasses, trees, and flowering plants yielded fruit. During the dry months the plateau dried up, lightning ignited fires, and rebirth came again with the rains. The savanna provided a nursery with a comfortable temperature, replete with fruit, nuts, and game.

There were fluctuations, however. Earthquakes and ever-changing rainfall patterns created jigsaws of local environments. When Earth entered a glacial period, the savanna became cooler and drier, with more grassland. During interglacial periods, the savanna was hotter and wetter, with more rain forest.

Climate is currently considered to be a key factor in evolutionary change. Apes changing to humans had to adapt to immense climatic swings. If climate had not changed as it did, if gene pools in particular places had not been exposed to particular pressures, especially coolness and dryness in the tropics, then our species probably would not have appeared as it did.

Earth entered its present condition of alternating glacial and interglacial periods only about 2 million years ago. The first ice sheets arose in Antarctica about 35 million years ago, after Earth took the previous 65 million years to cool off about 15 degrees Fahrenheit. Apparently in the last 2 million years the planet has gotten into a temperature range in which colder and hotter cycles can easily tip back and forth.

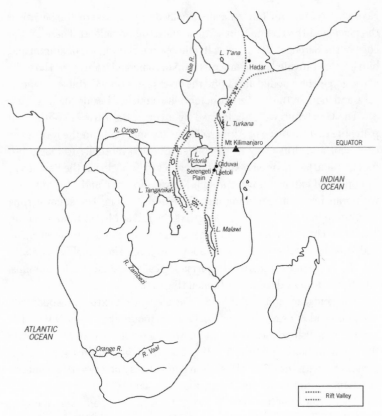

3.2   The Great Rift Valley of Eastern Africa

Over the last million years there have been about ten ice ages, or glacial periods, at intervals of roughly 100,000 years. The last one, called the Great Ice Age, began about 90,000 years ago and reached its glacial maximum about 20,000 years ago. In the last 10,000 warm years the temperature has been on average 1.8 to 5.4 degrees Fahrenheit warmer than it was during the preceding glacial periods, with periods of cooling.

What causes these fluctuations? They seem to be the result of tiny changes in the tilt of the Earth's axis, in its elliptical (oval) orbit about the sun, and in its wobble on its axis. Each of these has its own pattern—41,000 years for the tilt to vary from 21.39 degrees to 24.36 degrees and back, 95,800 years for the orbit to go from nearly circular to more elliptical back to nearly circular, and 26,000 years for the axis to work through one complete cone of wobble (precession). The effects of these three kinds of alteration are superimposed, sometimes intensifying each other, sometimes canceling each other out.

Other agents also cause fluctuations in climate—earthquakes, volcanoes, continental drift, changes in carbon percentages in the atmosphere, and hits of meteors and asteroids—not to mention that the magnetism of the poles sometimes reverses with a random pattern averaging every half million years or so. There have been 282 flips in the last 10 million years, known by the magnetism of the rocks at the bottom of the sea. The last reversal took place about 780,000 years ago, when *Homo erectus* was still learning to make stone tools. Scientists have currently noted that the strength of Earth's current magnetic field has waned 10 to 15 percent, and the deterioration seems to be accelerating, causing a debate over whether a field reversal, which typically takes 5,000 to 7,000 years, has begun.

From about 6 million years ago, through extremes of climatic change, the lineage of bipedal apes developed slowly and erratically. As many as twenty species of them once existed; now we are the only one left. The fossil evidence for this development is sketchy, fragile, and confusing. Many species coexisted simultaneously. Paleoanthropologists agree that a clearcut family chart cannot yet be constructed and maybe never can. By comparing the genome of humans with that of chimpanzees, scientists have identified a partial list of the genes that make people human. They include genes for hearing and speech, genes that wire the developing brain, genes for detecting odors and for shaping bones.

Specialists call the oldest group of bipedal apes *Australopithicus*, or southern ape. These creatures stood about 3 to 5 feet (1 to 1.5 meters) tall, with a head size like a chimpanzee. The oldest bones, dated about 4.4 million years ago, were found in 1992 at Afar, Ethiopia. The best-known *Australopithicus* is Lucy, whose less-than-half-complete skeleton was found in 1974 near Hadar, Ethiopia. She was named for the Beatles' song, "Lucy in the Sky with Diamonds," since the crews of excavators had this song playing as they worked. The bones at Hadar, parts of at least thirteen people, date from 3.2 million years ago.

Lucy was a north African ape that walked upright. She stood 3.5 feet tall, weighed less than 66 pounds (30 kilograms), was 19–21 years old, and had a pelvis like a modern woman but the face of a chimp. Her bones settled a long-standing controversy among anthropologists: Which did the human lineage develop first, large brains or bipedalism? The answer found in Lucy's bones is bipedalism. Lucy's skeleton showed that some great apes came down from trees, keeping their arboreal rotating arms and shoulders, and developed upright posture before their brains began to expand.

Another haunting image from the mists of our human development is a set of footprints found at Laetoli, Tanzania, in the late 1970s by excavators led by Mary Leakey. The footprints are those of two early bipeds who appeared to be walking across a field of ash from an erupting volcano. Their feet sank deep into the ash, which was damp from a light shower of rain. As the ash dried, the lime in it hardened. More ash fell into the prints, preserving them to be uncovered 3.6 million years later. What a find for humankind!

How did early great apes develop bipedalism? Specialists theorize that as apes in eastern Africa grew larger, they needed more food, which grew harder to obtain in trees, as forests were changing into grassland savannas. Probably the apes came down to look for food, then carried it back to their group. Being able to stand upright conferred the advantage of seeing farther, of carrying food and babies, and of freeing the arms and hands for other tasks. As legs became stronger and heavier, the body's center of gravity shifted downward, making an upright posture easier to maintain. Minor improvements may have begun to operate as a self-reinforcing system.

Several species of *Australopithicus* endured simultaneously until maybe half a million years ago, definitely complicating the lives of paleoanthropologists trying to sort this all out. Meanwhile, other species developed, and by about 2.5 million years ago the lineage *Homo* had appeared as a small-boned, larger brained ape. By 2 million years ago, *Homo habilus*, or handy ape person, had appeared. In these ape people, standing about four feet high, the brain had started its expansion, from 300 to 400 cubic centimeters in chimpanzees to 600 to 800 cubic centimeters in *habilus*. As hands were freed from walking and swinging, they began to work at creating stone tools, which helped to develop brains. Eyes strained to see farther, which also developed brains. Bigger-brained males must have chosen females with wider pelvises. Bigger-brained babies were born earlier than usual in the pregnancy in order to make it out through the birth canal, and they required prolonged care, calling for more adult interaction. The advantage of big brains showed up in the production of the first stone tools and in increased cooperation, although the ability to talk was still in the future. The hand-eye-brain-cooperation self-reinforcing loop had begun.

Handy ape people may have been the first tropical daytime hunters, or at least scavengers, possibly obtaining about 10 percent of their calories from meat. The whole question of how much meat early ape people ate is frought with controversy. Since the way mammals live is largely shaped by what they eat, this is a crucial matter for discussion. There is little real evidence, however—only inferences from structures of teeth.

By about 1.8 million years ago, along came *Homo erectus* (erect people), someone taller, as high as 5 feet 6 inches (1.67 meters), and bigger brained (900–1,100 cubic centimeters). Since the average modern brain size is 1,350 cubic centimeters, it seems to be time to omit the word "ape" and call this creature a person.

Homo erectus begins to seem quite familiar. They created wooden spears and chipped stone into beautiful hand axes. They may have hunted big game, which requires precision tools and elaborate social cooperation, although they probably had only the most rudimentary speech. Perhaps 20 percent of their calories came from meat. They established home bases and cared for dependent infants. They probably made the critical transition from the male and female hierarchies of chimp society to the conjugal bonds of modern men and women.

Erect people also learned to be unafraid of flames, probably instigating one of humanity's great leaps—the use of fire. They learned to conserve the embers of burning tree stumps, which had been lit by lightning, to produce their own fires. The advantages of this risky behavior proved enormously worthwhile. People could frighten off predators, set fires to drive prey into traps, cook and eat a wider range of food, preserve food for long periods, light up dark caves, and stay warm in colder weather. Bigger brains were really beginning to pay off.

Indeed, some argue that the preparing, cooking, and sociable eating of food are so central to the human experience that the culinary arts may have been a central component of how we became human. They certainly allowed people to eat more items and receive more nutrition from them. Cooking meat rather than eating it raw at the site of the hunt may explain both the relative lack of difference in size between males and females (as females got more to eat) and the tendency of couples to stay together longer than most other primates. Estimates of when people began using fire vary from 2 million to 300,000 years ago.[4]

With fire, erect people took another first step: some of them moved out of the comfort of warm Africa for the first time, taking their fire with them to protect them against the cold. This likely happened about 1.2 million to 700,000 years ago, during a warm, wet period when the Sahara Desert had enough rainfall to cross safely. Erect people probably crossed at the land bridge where Africa joins Asia at what is now Saudi Arabia. This movement should not be thought of as a migration but simply as small bands of hunter-gatherers moving in their quest for food. Eventually erect people moved into the Near East, Europe, parts of northern Asia, and tropical southern and southeastern Asia. They could not inhabit extremely cold places, like most of northern Eurasia. Erect people did not reach Australia and the Americas. The whole human world contained probably a few tens of thousands of people. But people, like other animals, undertook globe-trotting; the lineage *Homo* is as peripatetic as any other. Traveling at 10 miles (16.09 kilometers) a year, it takes less than 2,500 years to walk around the Earth. During the era of *Homo erectus* the saber-toothed tiger became extinct. Were people already having an effect on their environment?

A possible alternative scenario has people leaving Africa about 1.8 million years ago, then developing into *H. erectus* in Asia, and returning to Africa. The reality must have been a highly complex process of much population movement over time, with all kinds of local expansions and contractions.

Descendants of *Homo erectus*

The descendants of *H. erectus* can be grouped by three different locales: Neanderthals in Europe and the Mediterranean, *H. erectus* in eastern Asia, and *H. sapiens* arising somewhere in eastern or southern Africa. The official classification does not reflect this clearly, because Neanderthals were earlier thought to be a subgroup of *Homo sapiens*. Hence they were called *Homo sapiens neanderthalensis,* the designation they retain, even though they have been shown not to belong to the species of *Homo sapiens*. True *Homo sapiens* are officially called *Homo sapiens sapiens* to distinguish them from *Homo sapiens neanderthalensis.* For brevity I am using the terms Neanderthals and *Homo sapiens*.

Neanderthals are in the fossil record from about 130,000 to 28,000 years ago, originating before the start of the last ice age, about 90,000 years ago. They were the first humans to adapt successfully to life on the edge of an ice age world. More bones of Neanderthals have been found than of any other hominid group, including some thirty nearly complete skeletons. They are named for the Neander Valley skeleton found near Düsseldorf, Germany, in 1856, even though there had been earlier finds of this type.

The adaptation of Neanderthals to cold can be seen in their skeletons. Their bones are shorter and more compact than modern human bones, indicating a bulky, squat physique with heavy muscles and barrel chests in men, women, and children. Males were about five feet 6 inches (1.7 meters) tall and weighed about 155 pounds (70 kilograms), while females were close to 5 feet 2 inches (1.6 meters) and about 120 pounds (54 kilograms). Some features of the hip area indicate they did not walk exactly as we do.

Their brains were at least as large as our own, although shaped differently. Their skulls were long and low, like those of earlier humans, with a notable ridge above the eyes and a massive nasal opening, larger than in any humans before or since.

As toolmakers, Neanderthals did not change their designs over tens of thousands of years. From stone they made borers, scrapers, points, knives, and hand axes. They could hunt woolly mammoths, musk oxen, wolves, cave bears, wild horses, and reindeer, and lived mainly on a diet of hunted animal meat. They used wood but never perceived the possibilities in bone, antler, or ivory. There is no evidence of ornamentation until the very end of their existence, and no cave drawings.

Neanderthals definitely used fire. They scraped hides for clothing and shelter. They buried their dead—the first humans known to have done so. Bodies are found associated with tools, but there is no pattern of other burial goods or any clue to the possibility of ceremonies. Some skeletons show marks of illness or injury that occurred some time before death, so there must have been social care for disabled individuals.

What capacity for language the Neanderthals enjoyed is a topic for dispute. Anatomical reconstructions suggest that the larynx was positioned differently from that of modern humans, which must have limited the sounds Neanderthals could produce. Presumably their limited spoken language was extended by more gestures and facial and body language than we are accustomed to.

Modern geneticists have discovered that bones contain cells that do not vaporize immediately at death. Sometimes bits of DNA can be extracted from long-dead animals, so long as they are not dead too long. If a thousand years have passed since death, the chance of success in extracting DNA is about 70 percent. Yet in 1997 geneticists managed to extract a short sequence of DNA from Neanderthal bones from 30,000 years ago. The DNA suggested that Neanderthals were very different from a large selection of modern people and could not have been our ancestors. Neanderthals are now seen as a specialized form of *Homo erectus* adapted to extreme cold. (No DNA sample has yet been found for *Homo sapiens* of the same vintage.)[5]

As we shall see, by the time *Homo sapiens* finally arrived in Europe from its origins in southeastern Africa, Neanderthals were wiped out. It seems that the kind of people who evolved in Europe (e.g., Neanderthals) proved less fit than the kind who evolved in Africa. This fact could not be unraveled or assimilated by European minds until well into the 1960s and 1970s. Before that time, racist thinking, combined with the lack of excavations in Africa and poor dating techniques, prevented recognition of what now seems to be the accurate story line.

One example must suffice. In 1912 a find was announced of a skull of a large-brained hominid that had been located in the Piltdown gravels of Sussex, England. Piltdown man was taken in the European and American scientific communities as proof that the first large-brained human ancestor had evolved in England. The Piltdown skull became the standard against which other skeletal evidence was judged and found wanting.

Forty years after Piltdown man was found, it was proved to be a fake—an ingenious combination of fragments of a modern human skull and an orangutan jawbone, both doctored to look ancient. The perpetrator of this hoax has never been identified; the suspects include the anatomist who first commented on the remains, the amateur archaeologist who found them, the museum curator with a grudge against the archaeologist, and even Sir Arthur Conan Doyle, the creator of Sherlock Holmes and a friend of the archaeologist. This joke calls into question the credibility of the entire scientific enterprise. In the end, however, European archaeologists were able to defend their creed by unmasking this fraud, though it took them forty years to do so.[6]

In eastern Asia, *Homo erectus* was the first human species to arrive; there, erect people developed distinctive adaptations to the forest environments of tropical and temperate Asia. Forests rather than grasslands meant that people had to keep moving in order to find fruits and nuts. Instead of stone for their tools, they used bamboo and wood, raw materials that are not preserved in ancient sites. These forest cultures flourished and evolved slowly over hundreds of thousands of years, quite independently, it seems, of changes in humankind in Africa and Europe. *Homo erectus* seems to have lasted several hundred thousand years longer in Asia than in Europe and Africa. In the unforgettable words of linguist Derek Bickerton, *H. erectus* in northern China "sat for 0.3 million years in drafty, smoky caves of Zhoukoudian, cooking bats over smoldering embers and waiting for the caves to fill up with their own garbage."[7]

Homo sapiens Inhabits the World

So we come, at last, to us, modern human beings. Once again we find ourselves somewhere in eastern Africa where, sometime between 250,000 and 130,000 years ago, African descendants of *Homo erectus* mutated one more time into a fitter species, *Homo sapiens,* the last speciation event so far in the human line.

Homo sapiens was tall and thin, not robust like the Neanderthals. *Homo sapiens* had no brow ridges and a larger, higher forehead and cranium. Even though its brain capacity was smaller than the Neanderthals', it was shaped quite differently.

Homo sapiens seems to have been the first group of hominids to develop fully articulate speech. Full fluency enabled *Homo sapiens* to develop syntax and the distinctive human qualities of abstract, rational, symbolic thought. The hand-eye-brain-speech reinforcing loop was in full gear.

Since there are only two anatomical ways to study the evolutionary development of human speech, our knowledge of it is one of the least developed areas of the human story. One can study the area of the brain that controls speech, called Broca's area, whose size and shape can be deduced from endocasts taken of the inside of skulls, or one can study the development of the larynx and pharynx from the bones of the throat.

Broca's area of the brain appears also to control precise hand movements. Speaking requires precise movements of the tongue, analogous to those of the hand. Research has shown that people who have suffered brain damage that impairs their ability to produce and understand language also cannot perform sequences of precise hand movements. Autistic children who learn sign language can sometimes become able to speak. Theorists believe that as early people developed fine finger movements, they also developed the area of the brain that enabled them to sequence words and develop syntax. Gesture moves in a continuum into speech.

Peculiar to humans is the position of the larynx. In most other animals it sits high in the throat and serves as a valve to separate the air flowing to the lungs and the liquid flowing to the esophagus. Other animals can drink and breathe simultaneously; we cannot. Our larynx has repositioned itself halfway down our throats, as shown by the position of the Adam's apple in the adult male. This leaves a space at the back of the nose and the top of the throat that serves as a sound chamber, giving us a resonance that no other species has. This resonance, combined with the extreme dexterity of our tongue and lips, gives us a verbal fluency that is comparable to our amazing manual dexterity. Children reenact the evolutionary history of the larynx; as they grow, it drops from the top of the throat to its full human position by about fourteen years of age.[8]

Whether the location of the larynx was fully low-slung in Neanderthals is the subject of much debate. Most experts believe that in them the larynx was in an intermediate position, about like that of an eight-year-old child. All agree that by the time the noticeable takeoff in human culture occurred, at about 30,000 years ago, the larynx was in its modern position, and humans had mastered speaking.

Over time, *Homo erectus* in Africa gradually turned into *Homo sapiens*. Using their superior brains and language abilities, they gradually eclipsed other human species in Africa and achieved a population of maybe 50,000 by about 100,000 years ago. Some of them managed to move—in a short window of opportunity provided by Earth's climate—out of the tropical savannas into the area of the eastern Mediterranean that is now Israel, Palestine, Syria, and Lebanon. Then, about 90,000 years ago, Earth tipped back into a cold glacial period in which the Sahara Desert rapidly dried up, preventing further human crossing until warmer, wetter times.

Homo sapiens does not seem to have reached Europe until about 60,000 to 40,000 years ago, even though they were in the eastern Mediterranean by 90,000 years ago. This poses the question: Why did it take so long for them to reach Europe?

Speculation has it that *Homo sapiens* needed time to adapt to weather colder than that in the warm savannas. People stayed in the eastern Mediterranean learning the skills required by the coldness—creating better clothing and shelter, and developing better hunting techniques since fruit and nuts were scarcer. During a short period of warming weather, 50,000 to 40,000 years ago, *Homo sapiens* moved into southern Europe, (ice sheets still covered northern Europe). When cold weather returned, these *Homo sapiens*, now known as Cro-Magnon people, adapted, not through physical changes as Neanderthals did, but by using advanced cultural skills.

This poses a fascinating question: What were the relations between Neanderthals and *Homo sapiens*? We know that they interacted in the Middle East and in central and western Europe, but we can only imagine how. Since they are classified as subspecies of the same species, earlier experts assumed that they could mate together. But modern geneticists doubt that *Homo sapiens* mixed its genes with Neanderthals. There may have been warfare between the two groups. Or there may simply have been a different mortality rate, due to skills in adapting. Supposing a difference of 1 percent in mortality rate, Neanderthals might have become extinct in thirty generations, within a mere thousand years, or a millennium. However it happened, by 32,000 to 34,000 years ago, Cro-Magnon people had become the sole hominids of Europe.

As some groups of *Homo sapiens* spread westward around the Mediterranean Sea into southern Europe, presumably other groups moved eastward into southeastern Asia. As yet there is no fossil evidence to substantiate this. We do not know when *Homo sapiens* colonized southeastern Asia, the Indonesian Islands, and the continent known to geologists as Sahul—a combination of New Guinea, Australia, and the shelf between them, now flooded but then above water as frozen glaciers reduced the sea level.

The colonization of Sahul by early people constitutes the first instance of seafaring by human beings. At the height of glaciation 20,000 years ago, the distance between the mainland and Sahul was about 62 miles (100 kilometers) of open water. It is known that people crossed this water earlier than this, with less glaciation and higher sea levels, when the distance on open sea must have exceeded those 62 miles.

Which people achieved this amazing feat—descendants of *Homo erectus* or *Homo sapiens* arriving from Africa? No one knows. It seems likely that the initial settlement was the result of both accidental and deliberate voyaging by tiny numbers of people in bamboo rafts. These trips, lasting at least seven days, must have taken place over thousands of years. Hunter-gatherers are known to have been living throughout much of New Guinea and Australia at least by 40,000 years ago; possibly they were there as early as 50,000 to 60,000 years ago. These early settlers in New Guinea and Australia were the first people to build boats capable of traveling over considerable stretches of open sea.

The settlement of central Asia, China, and Siberia was a complex process, not yet fully understood. Recent dental research has shown connections between teeth in northeastern Asia and those in southern China, rather than to those in either Europe or southeastern Asia. The differences in teeth are so striking that experts believe that the people of northeastern Asia were a group of *Homo sapiens* separate from those in southeastern Asia or Europe. These northeastern Asians may have eventually crossed over into the Americas and completed the peopling of the world.

During the Great Ice Age, the sea level fell far below modern levels. Land existed between Siberia and Alaska, a now-vanished continent called Beringia. This land bridge was at its maximum extent at the most intense glaciation about 50,000 years ago and again at 20,000 years ago. When the melting of the glaciers commenced, the rising seas began to cover Beringia, submerging it completely by 12,000 years ago. Any time between 90,000 and 12,000 years ago, it would have been possible for Ice Age hunter-gatherers to cross Beringia to the Americas without using any watercraft.

But when and how did they? Most experts agree on Beringia as the route. The earliest unequivocal archaeological evidence of settlement comes from the Clovis site in New Mexico, dated at 13,600 years ago. There are hints of earlier settlement, possibly back to 30,000 years ago.

However it happened, the settlement of the Americas was the culminating development in the slow expansion of modern humans from Africa—first into other tropical and temperate climates, then into the near glacial environments across the north of Eurasia, then to the new continents. By 11,000 years ago, hunting and gathering people occupied every corner of the Americas. Like their relatives around the world, they were inventive and ingenious, as they had to be to survive. They adapted to local environments in highly diverse ways, leading to the dazzling array of cultures that greeted people from Europe who arrived 12,000 years later.

The final steps in occupying the last inhabited islands of the world were taken by Polynesians, who sailed to Tonga and Samoa about 3,000 years ago; to the Marquesas, Rapa Nui (Easter Island), and Hawaii by 1,500 years ago; and to New Zealand about 1,200 years ago. People from Indonesia settled Madagascar about 1,200 years ago (Fig. 3.3).

In its expansion from Africa, *Homo sapiens* remained one species around the world. Even though it has been 100,000 to 200,000 years since modern people left Africa, *Homo sapiens* has not broken up into separate species, unlike chimpanzees, who broke into two distinct subspecies about 2 million years ago, separated by the Congo River. There really has not been enough time, genetically speaking, for humans to split. In addition, human populations have maintained contact with each other, over vast distances of time and space. The Great Ice Age must have assisted in this— the glaciers froze so much water that sea levels remained low and continents were connected, enabling at least some members of *Homo sapiens* to roam freely, mating back and forth at the edges of separate populations. Perhaps it was early members of trekking clubs who kept our species intact.

In summary, full human beings, the descendants of tropical arboreal apes, first made their appearance in eastern Africa between 100,000 and 190,000 years ago. They moved out from Africa to inhabit the Earth and survived some of its harshest climates to become the dominant life-form in every land they entered. In retrospect, it is astonishing how recent the current human is—only 100,000 to 200,000 years old—when our immediate ancestor, *Homo erectus*, took 1.4 million years to turn into *Homo sapiens*, not to mention the 3 to 4 million years it took before that, back to the common ancestor, for our divergence from apes. Indeed our species may still be

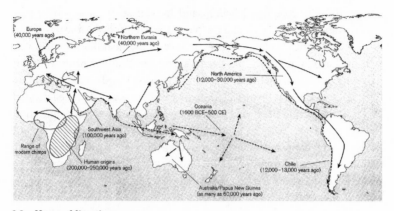

3.3   Human Migrations
(*Source:* David Christian, 2004, *Maps of Time: An Introduction to Big History,* Berkeley, CA: University of California Press, 193.)

in its childhood, if we have anything like the few million years that most species enjoy.

From the story of how humans became the current dominant life-form, two main ideas seem worth emphasizing. One is how humans are inherently part of all life. We are vitally connected to the deep rhythms of the Earth and to all of its living forms. Our religions, psychologies, and philosophies have tended, for several centuries at least, to obscure and downplay our biological connections to Earth, as do our urban living arrangements, but in recent years awareness of our connection to all of life has been increasing among Western people. People close to the Earth, of course, have never lost this awareness.

Another main idea is that conditions on our Earth over time are never the same. Even though on a daily basis Earth seems quite permanent, nothing is further from the truth. A combination of forces on Earth produce extraordinary complexity and unpredictability, and forces that seem to be acting smoothly may result in sudden changes. Looking at the big picture, we see that we live in the gaps between what we consider catastrophes. We are learning to live with the awareness of these long-term changes while at the same time making the assumption of everyday permanence that is appropriate for conducting our short-term lives.

After three chapters, our story can be summarized thus: Our universe began 13.7 billion years ago as a speck of incomprehensible energy, which burst into expansion that continues still. After sufficient cooling, matter appeared as hydrogen and helium and formed stars, in which heavier atoms were created. Some stars exploded as supernovas, scattering the heavier elements, which formed new star systems, including our solar system and Earth. With the help of energy sources, such as ultraviolet rays and lightning, the chemical building blocks of life developed on Earth, leading eventually to the first living cell 3.5 to 4 billion years ago. This cell divided and multiplied, and life has been evolving from it ever since. A mutation about 6 million years ago started the development of chimpanzees toward humans, who appeared as a species a mere 200,000 to 100,000 years ago, dominated other human types by 30,000 years ago, and inhabited the planet by 13,000 years ago (Fig. 3.4).

Unanswered Questions

1. Where and when did *Homo sapiens* first appear?

In explaining the origins of archaic humans, scientists agree that *Homo erectus* evolved in Africa and spread out from there about a million years ago. Until recently scientists were still divided into two camps about the origin of modern humans (*Homo sapiens*). One camp argues that our most recent ancestors evolved independently and in parallel in different parts of the world. This hypothesis is called the Candelabra theory—each branch of human development is like a branch of a candelabra. The other camp, now in the majority, presents the story as it was developed in this chapter, that modern humans evolved in Africa and radiated out from there to the rest of the world. This hypothesis is called Noah's Ark (we were once all in one boat), Out of Africa, or Garden of Eden.

According to the Candelabra theory, modern humans emerged in many places and diverged genetically at least 700,000 years ago and probably earlier. Noah's Ark theorists argue that modern humans emerged in Africa 100,000 to 200,000 years ago, then spread out, with a much more recent genetic divergence. The Candelabra theory enjoyed popularity when most of the fossil specimens came from Europe, the Near East, and Asia. When fossils began to be found in Africa in the 1970s, many scientists shifted to the Noah's Ark theory. Most of the recent evidence of various types points to a late divergence in Africa, but it is not yet fully conclusive.[9]

The two theories have very different implications for the anatomical diversity in modern geographic populations. Noah's Ark theorists say that differences in skin color, hair form, and build are superficial, recent adapta-

3.4   The Big Picture
(*Source:* W. J. Howard, 1991, *Life's Beginnings*, Coos Bay: Coast Publishing, 84–85.)

tions to different environments. Candelabra theorists say these genetic differences go back a million years.

2. How can the findings of religion and science be reconciled?

Some people in the Judeo-Christian world, as do people in other religious traditions, reject the findings of science and continue to believe that God created the world as it is just a few thousand years ago. These people are known as "creationists," and their position is called "young earth creationism." There are other creationist positions. "Old earth creationists" accept modern geology and astrophysics, but reject the findings of biology, specifically evolution. Other creationists accept some of evolution, but not the continuity between extremely different kinds of creatures, specifically that between humans and apes. Many Africans, being familiar with apes, believed that people were descended from them, but that idea was not acceptable to Christianity or to Islam.

According to Gallup polls, in 1997, 44 percent of Americans believed that God created humans pretty much in their present form in the last 10,000 years, while only 10 percent affirmed evolution without God's participation. The rest held that God guided the evolutionary process in some way.[10] Most Americans try to synthesize evolution with the existence of a personal god.

Many leading scientists, such as Brian Goodwin, Richard Lewontin, and Richard Dawkins, do not believe this is possible. They see no progress or direction in evolution, but rather a set of improvisations and haphazard events, an unbridled creativity of life, a dance exploring the space of possibilities.

Other scientists work out of their religious background and/or address themselves to religious people. Examples of such books are Brian Swimme and Thomas Berry, *The Universe Story: From the Primordial Flaring Forth to the Ecozoic Era*; Ursula Goodenough, *The Sacred Depths of Nature*; Fritjóf Capra and David Steindl-Rast, *Belonging to the Universe: Explorations on the Frontiers of Science and Spirituality*; and Edward O. Wilson, *The Creation: An Appeal to Save Life on Earth.* A religious philosopher looking at the whole story from a naturalist point of view is Loyal Rue, *Everybody's Story: Wising Up to the Epic of Evolution.*

4

Advanced Hunting and Gathering
(35,000–10,000 Years Ago)

Now that humans have entered our story, it must slow down percepti-
bly and take a closer look at this strange, hairless ape with a voice box
and a big brain: us. The last 30,000 years, in the context of the time already
elapsed, seem to be contemporaneous with our time. After all, people
30,000 years ago are only 1,200 generations removed from us. At twenty-
five years per generation, four generations cover a hundred years, forty generations cover 1,000 years, 400 generations cover 10,000 years, and 1,200
generations cover 30,000.

As we have seen, the modern mind and modern behaviors evolved spo-
radically in Africa from about 200,000 years ago. Gradually people developed their symbolic language and passed down their collective learning,
until by about 35,000 years ago people were producing cave paintings, carvings, figurines, grave goods, ornamentation, and presumably full symbolic
speech. The complexity, refinement, and symbolic representation found
from 35,000 to 12,000 years ago, plus the successful adaptation of humans
to all parts of the world, have convinced prehistorians that these people are
the same as us, with the full speech and brain power of modern people.

What was life like for these advanced hunter-gatherers? I am calling
them advanced because they had gradually built up a hunting and gathering life over 100,000 to 200,000 years until, in the period under scru-
tiny, they were able to advance considerably the complexity of their
activities.[1]

The Hunting and Gathering Life

The lives of hunter-gatherers must have varied widely, since we know that they lived in almost all of Earth's climates and skillfully adapted to their particular environment. Nevertheless, we can consider the general characteristics of people living as hunter-gatherers, whether in the Arctic or the Amazon, the deserts of Australia or of South Africa.

Hunter-gatherers, sometimes called foragers, lived in small groups large enough to defend themselves and divide tasks yet small enough not to exhaust food supplies within walking distance. Groups varied in size, depending on the availability of food, varying from ten to twenty to perhaps about sixty to a hundred people. Occasionally groups would converge, but they could not feed themselves for long. Historians imagine that before 20,000 years ago, there were probably never more than 500 people gathered at one time.[2]

These groups led a primarily nomadic life, moving from site to site as animals moved or plant foods were consumed. The pattern of moving depended on the local environment. Groups might stay for several months during the summer, but move more frequently during spring and fall. Winters would find them holing up in caves near animal supplies. A few groups, like those on the Pacific Coast of North America, could form permanent settlements because of the constant abundance of salmon and other seafood.

Diets probably varied seasonally, except in extreme climates like the Arctic areas, where the Inuits lived solely on hunted meat. Hunted meat, or scavenged meat killed by other animals, may have constituted from 10 percent to almost the entire diet of hunter-gatherers, and it probably usually varied seasonally. Scavenging may have predominated until about 30,000 years ago, when consistent hunting may have begun with improved tools.

It is difficult to assess the quality of foragers' diets, but recent evidence suggests that it may have been much better than formerly supposed, depending again on luck and locale. Fruit and nuts, combined with feasts of meat, may have provided nutritious satisfaction. Infectious disease was not prevalent, since people moved before pollution and contamination built up. Since many deaths in infancy, accidents, and warfare occurred, the average lifespan is thought to have been about thirty years. By this period, however, at least some people were living into their sixties.

The creativity of hunter-gatherers revealed itself in their shelters. We know they used south-facing caves when available. They used the bones of large animals to construct dwellings. For example, at Mezhirich in the Ukraine, dated about 15,000 BCE, woolly mammoth bones were used. Woolly mammoths weighed about 10,000 pounds, and one structure contained bones from ninety-five of them. Less durable materials must have been used in other places—animal skins, tree branches, stones, rushes, and mud. Since in many areas life required frequent moving, in those places the most desirable shelter materials must have been lightweight and portable.

Since stones and bones are about all that remains after 10,000 to 30,000 years, the lives of hunter-gatherers would be impossible to imagine except for the fact that a few people today are still living that kind of life. There are not many, and they are diminishing rapidly under pressures from the non-hunting-gathering world. Still, they are a testament to their way of life.

Can we rely on the way current hunter-gatherers live to tell us about their lives 10,000 to 30,000 years ago? Certainly there are some reasons why their modern ways might be different. For one, they have evolved and developed over the ensuing years just like people living agricultural or industrialized lives, and hence their lives are not likely to be just as they were 30,000 years ago. In addition, foragers are now constricted to areas smaller than they need, often in extreme environments that no other group wants to inhabit. Now they are struggling for existence, whereas in earlier times they roamed larger and more abundant areas. Finally, very few current foragers are entirely untouched by modern technology or politics.

Nevertheless, whether in the Arctic, the rain forests, or the deserts, current hunter-gatherers still share so many characteristics with their predecessors that anthropologists feel they can generalize from them to help construct what life must have been like 10,000 to 30,000 years ago.[3]

From contemporary hunter-gatherers we know that men almost always do the hunting, while women do the gathering of plants and small animals. The fishing and small-game hunting frequently are shared. Men usually create the tools associated with hunting—spears, bows, and arrows—while women usually create containers for collecting and cooking food, such as baskets, slings, and pots, and the tools for making clothes, such as scrapers, needles, and sinew. Social structure is simple and egalitarian, without strata or hierarchies, other than those distinctions of age, gender, kinship, and personal achievement that seem present in any human society. People share food by necessity.[4]

The remains of tools that have been excavated from the 20,000 years before the development of agriculture show a breathtaking advance in quality, quantity, and creativity. Before this time, stone tools tended to be large, mainly hand axes and flakes from a prepared core. After this, stone tools tended to be thin, double-edged blades, and eventually these thin blades were used as points or projectiles. People began to use materials other than stone—mammoth ivory, bone, and antler—for making tools, some quite complex, such as barbed harpoons and fishhooks. By 23,000 years ago people had invented the bow and arrow, making hunting much easier. By 23,000 years ago people had invented the spear thrower—a handle a foot or so long, often made of deer antler decorated with animal forms, with a hook on one end on which to fit the spear. In the cave at Lascaux, France, dating from 17,000 to 20,000 years ago, archaeologists found an imprint in clay of a three-stranded rope. People who could make cords could produce nets, traps, and snares.

The domestic arts began. People created stone-lined fireplaces and lamps to light their caves, most often small slabs of limestone hollowed in the center to hold animal oil. Sewing needles, mostly made of bone or ivory with eyes for sinew, appeared about 20,000 years ago. With needles, animal furs could be sewn tightly together for warmth. Bracelets, necklaces, and beads were created from mammoth teeth, fangs, shells, and bones. Women as well as men created tools; we can only guess who created which tools.

The population during hunting-gathering times stayed relatively stable. Many infants died, and probably their births were spaced every four or five years. Women would not have been able to carry two babies while gathering or moving. Part of this spacing occurred naturally as women nursed their infants for several years, since other milk or cereal was not available; nursing tends to suppress ovulation. Other means of birth control were probably practiced: infanticide, especially of one twin; herbal ways to induce abortion or to prevent conception; and sexual abstinence. Low fertility also resulted due to food shortages.

Children played while women prepared food; soon enough children helped to collect seeds and fruits and to hunt lizards and frogs. With a little luck from weather and locale, this life could have been a relatively affluent one, with all the basic needs provided by a few hours of work a day, leaving much time for socializing, grooming, and relaxing.

The artifact that most suggests the awakening of a more complex human consciousness is paintings found in caves. These are found around the world, but have been best preserved in deep limestone caves in southwestern France and northeastern Spain, on either side of the Pyrenees. The first one was discovered at Altamira, Spain, in 1879. Since then more than 200 caves with paintings and engravings have been found in this area, where people retreated during the height of glaciation some 25,000 to

20,000 years ago.* (Large herds of reindeer and red deer made a moderately dense human population possible without the need for long migrations to find food.) The best-known cave paintings date from about 28,000 to 13,000 years ago, ending as warmer weather increased food supplies and made relating to animals less important.

The paintings that people made on cave walls predominantly depict animals, specifically deer, bison, horses, and aurochs. Human forms are numerous, including handprints and images of male, and especially of female, sexual organs. Recent analysis of the handprints suggests that they were mostly the hands of adolescent males. Whether or not the art resulted from any religious or magical motivations remains entirely unknown.[5]

The first unmistakable human and animal figures date to 30,000 to 32,000 years ago, as does the earliest evidence of music—a wind instrument made from a bone with four holes on one side and two on the other. A cave called Gargas in the French Pyrenees, dated 23,000 to 26,000 years ago, features more than 200 human handprints. These are "negative" prints, made by brushing or blowing pigment through a tube around a hand held flat against a rock surface. All except ten hands represented have fingers missing. How are we to interpret the missing fingers? Were they the result of ritual mutilation, or disease, infection, accident, or frostbite, or were the fingers deliberately held down against the palm in some kind of code? How can we know? Handprints on cave walls have been found around the world—in Australia, Brazil, and California.[6]

From about 25,000 to 23,000 years ago the most distinctive art forms found in Europe were the female statuettes that have come to be known as Venus figures. These little statues were usually carved from stone or mammoth ivory; fewer were formed from clay. Most were female, some with exaggerated proportions, others not. Some were male or sexless. Representation of stylized vulvas and phalluses are among the found objects.

Some archaeologists and historians believe that the Venus figures indicate a widespread worship of fertility goddesses and a time when women were regarded with respect and awe for their power to produce new life before patriarchy set in.[7]

No evidence confirms or refutes this interpretation; we are left guessing. The figures may have represented: a mother goddess or a goddess of

---

* The people in this region today, the Basque people, are genetically distinct from all other Europeans in their high incidence of Rh-negative blood. Their language is also distinct, suggesting that they may be the descendants of the first *Homo sapiens* in Europe who vanquished Neanderthals before other *Homo sapiens* people moved in from the Middle East.

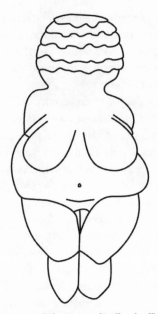

4.1   A Female Figurine Known as "The Venus of Willendorf"
The figurine is carved in limestone, 4½ inches high (11cm), and about 25,000 years old.

fertility, a spirit invoked to protect newly built homes and hearths, a teaching aid in initiation ceremonies, or a talisman for fertility. Other possible interpretations: children used the figures as dolls or adolescent men used them as sex fantasy figures (Fig. 4.1).

Several objections have been made to interpreting these figurines as fertility goddesses. The belief systems of today's foragers center on general spirits and forces, not on personified gods and goddesses. Calling them Venus figures makes an analogy to a Roman goddess, the product of a way of thinking of a different culture, which individualized the deities. Today's foragers, in addition, are more concerned with limiting their population than with increasing it; perhaps this is partly a function of the diminished area available for modern foraging, but it also fits with our interpretation of early foragers' apparently stable population. Whatever interpretation of Venus figures one chooses, the dominance of female representation over male must somehow be significant.[8]

Today in postindustrial society, with its extreme stresses, it is easy to idealize the life of hunter-gatherers. The joy of living outdoors among the animals shines out from the images on walls in the recesses of caves, touching a lost chord in us. Being able to gather necessities by working only several hours a day, with no fixed schedule or deadline and surrounded by friends and family, certainly seems appealing.

Yet life must also have been uncertain and frightening. People lived in flimsy shelters among huge predatory beasts. Leopards snatched children from caves at night. Weather and food supplies could never have been certain. Death often came suddenly and unexpectedly. Foragers sought consolation and emotional expression in feasts, art, music, rituals, and gatherings—exactly as we ourselves do today.

What Did Hunter-Gatherers Speak?

Experts feel certain that hunter-gatherers were speaking with each other in some kind of language. As discussed earlier, Neanderthals probably did not have full human speech, but *Homo sapiens* did. The genetic mutation that switched us to the new species now seems to have been a neural change in the brain that gave us the use of grammar and syntax. (Using syntax means replacing the random stringing together of words, seen in toddlers, with arrangements of words in hierarchical structures, using substructures like clauses signaled by abstract markers such as "because," "although," "unless," and "since.") This ability proved such an advantage that it enabled *Homo sapiens* to prevail over all other hominids.[9]

Just how and when full symbolic speech developed is still not understood. Some think that something special happened in the networking of the human brain or in the structure of the larynx and tongue—rather suddenly about 60,000 to 40,000 years ago—that enabled full speech to arise.[10] Others believe that full symbolic speech happened earlier and more gradually, developing after the genetic switch to *Homo sapiens* and involving many other factors, perhaps primarily the collective learning that accrued in tandem with symbolic speech. Recent theories from evidence in Africa suggest this gradual development, which showed up suddenly in Western Europe when *Homo sapiens* migrated there.[11]

Some linguists believe there must have been one original human language shared by the first group of *Homo sapiens* in Africa. Most, however, do not believe that this language can ever be reconstructed, because too much time has elapsed since it was spoken. Reliable reconstructions go back only several thousand years.

Yet some linguists continue to look for clues to an original language. Some are guessing that the original language probably contained click sounds—consonants made by sucking the tongue down from the roof of the mouth. Click sounds require active use of the mouth and tongue, especially when making several in a row. The only remaining languages with clicks (usually four to five different ones per language) occur in southern Africa, used by those groups that genetically have been shown to go back nearly to the first group of *Homo sapiens*.[12]

Another piece of evidence for an early common language is the fact that a group of stars in the Taurus constellation is called the Seven Sisters by widely dispersed groups: native North Americans, Siberian people, and Australian people. Each group has its own words for the constellation, but they all translate as "Seven Sisters." (English-speaking people call the sisters "the Pleiades," from Greek mythology, in which Atlas and Pleione had seven daughters whom Zeus placed in the stars.) This seems unlikely to have happened by coincidence; the various people must have carried this term from ancestors common to all of them before they split up at least 60,000 years ago to inhabit different areas of the planet.[13]

Some linguists have traced about 500 words they believe belong to a language called "Nostratic," thought to have been spoken by hunter-gatherers across the Middle East, the Ural and Caucasus mountains, central Asia, India, and northern Africa sometime between 12,000 to 20,000 years ago. These linguists focus on the words that seem to have the most stable meanings—that is, they are seldom or never replaced by other words with the same meanings. Twenty-three of the most stable meanings are the following: I/me, two, you (singular), who/what, tongue, name, eye, heart, tooth, no/not, fingernail/toenail, louse/nit, tear (drop), water, dead, hand, night, blood, horn (of an animal), sun, ear, and salt.[14]

Language used to be seen as an artifact of culture, as in this statement: *Homo sapiens* developed a large brain and with it created language. This concept belonged to the paradigm dominant in the social sciences in the first seventy years of the twentieth century, when experts believed that humans were characterized by culture and that most of our behavior was driven by what we learned from the culture of our families and our communities.

In the past thirty years, however, that paradigm has shifted as biologists, and social scientists working with them, have stressed the biological determinants of human behavior. Physical nature, interacting with culture, is now seen as the fundamental determinant of human behavior. The remarkable human ability for language is seen as resulting from the structure of our brains, prewired for language acquisition. As a consequence, human infants acquire the language provided for them by culture with a speed and fluency incompatible with any process of trial and error.

As long as experts were looking at culture as the determinant of human behavior, they noticed the exotic differences among people. Now that they are looking at physical nature, they are finding similarities among diverse cultures. They are discussing human universals—traits shared by all individuals, all societies, all cultures, or all languages. Some human universals are inherent in human biology, while others are cultural conventions that have come to have universal distribution. The most obvious example of a human universal is complex, symbolic language; that seems to be the distinguishing feature of humans. Other examples are: all people have some form of shelter, are not solitary dwellers, have patterns of socialization, kinship, and division of labor, and distribute prestige differentially. All people have private sex and communal eating. In all cultures, men predominate in the public political sphere. People cooperate greatly and have more conflict than they wish. They distinguish right from wrong, carry out ritual, song, and dance, and mourn their dead. These are human universals.[15]

Another human universal, of course, is consciousness, or each person's awareness of self. How far back in evolutionary history does this extend? What other animals share it? When did full human consciousness appear?

Consciousness seems to come into play when a hierarchy of neural systems in the brain transmits signals at a high degree of complexity. When the signals become complex enough, the creature crosses the threshold into continuous conscious experience. Presumably, the human level of neural complexity increased gradually over time from that of our chimpanzee ancestors to the full consciousness that seems to have been present in humans 40,000 years ago.[16]

The Rising Seas

Strange as it may seem, human beings moved out into the world and occupied most of it during a period of intense glaciation, when the winters were much harsher than they are today. The Great Ice Age, lasting from 90,000 to 17,000 years ago and peaking at about 20,000 years ago, coincided with human colonization of the planet.

Even stranger, whereas the onset of glaciation had been gradual, the period of thaw proved swift, lasting no more than 5,000 to 7,000 years. About 17,000 years ago the change in climate began, a period of global warming in which the world's flora and fauna, including people, had to adapt quickly or perish.

By 14,000 to 11,000 years ago, as temperatures rose, the ice melted rapidly. The rising seas changed the human landscape dramatically, disconnecting people from areas formerly accessible and submerging up to 40 percent of the coastline. The land bridge between Asia and the Americas was submerged, forming the Bering Strait. The land connecting England to the continent sank under water, now called the English Channel. The sea rose over the land connecting Spain to Africa, forming the Strait of Gibraltar. Sri Lanka was separated from India, and the Philippines and Taiwan from Korea.

By 12,000 years ago the waters of Lake Victoria began flowing into the Nile River, creating the longest river in the world. Around the globe other large rivers formed—the Ganges, the Yellow, the Indus, the Tigris and Euphrates—and the flooding of those rivers deposited the silt that would permit the agriculture that would sustain the later cradles of civilization. Patterns of rainfall and vegetation changed rapidly. By 10,000 years ago the seas had risen 400 feet (140 meters) from glacial times.

The flooding continued. In about 5600 BCE the Mediterranean Sea rose so high that it crashed with great violence through the land bridge joining Turkey to Bulgaria, creating the Bosphorus Strait. The seawater from the Mediterranean Sea transformed a small freshwater lake, Lake Euxine, into the vast saltwater Black Sea. Displaced people appeared in various places— Hungary, Slovakia, and Iraq—as evidenced by linguistic analysis. This astonishing flood became seared into the memory of its survivors as the myth of the world flood; accounts of floods are included in about 500 of the world's mythologies.[17]

Africa enjoyed favored conditions during the period of the rising seas. Refugees from flooded areas poured into the southern half of Africa. Even northern Africa became attractive, as rainfall increased in the Sahara, creating lakes and marshes; the moistness continued until sometime after 5,000 years ago, when the deserts again began to widen.

During this period of global warming many large mammals all over the world became extinct. Many of them, like the woolly mammoth, woolly rhinoceros, mastodon, and steppe bison, had been favorite prey for human hunters for tens of thousands of years. Indeed, overhunting may have accelerated the extinction of these mammals; the issue is much debated. Only on the Great Plains of North America did a way of life based on big-game hunting continue to flourish, thanks to the survival of the North American bison.

Humans adapted to the climatic changes by broadening their diets to include more small animals, plant foods, sea mammals, and shellfish, as well as fish. These food sources increased in the warmer circumstances.

The earliest farming techniques—managing the environment, corralling animals, pruning and protecting plants—may have developed before the glaciers thawed but certainly they accelerated as the global climate warmed.[18]

The increasing complexity of obtaining food led to increasing complexity in human social organization. Populations increased. Both laid the foundations for the most significant human adaptation to global warming: the development of agriculture as people settled down to larger-scale farming.

During previous hunting-gathering times the human population had stayed relatively stable. Estimates put the world population in 28,000 BCE at several hundred thousand people. By 10,000 BCE the world population had grown to an estimated 6 million, the result of ice thaws, cultural advances, and collective learning by humans as they shared their achievements in social networks and passed them on to their children, enabling them to reproduce more successfully.

Genetic Drift and Adaptation

By the time our story reaches the beginning of agriculture, about 10,000 years ago, modern people had been living in diverse climates all over the planet for some 50,000 years, and they had existed as a species for some 200,000 years. To what extent did human beings change during this period of time?

Today human beings share about 90 percent of their DNA with the rest of the living world and approximately 98.4 percent with our closest nonhuman relatives, the chimpanzees. That leaves only about 1.6 percent that makes us human. The overwhelming part of this we all share. Tiny bits vary, causing some internal differences and some external differences in skin, hair and eye color, hair form, and facial shape. These tiny genetic differences vary by populations and cannot be divided into neat subcategories of people. The distribution of one difference does not match with the distribution of another difference. There are probabilities of sharing certain genes, but there is no genetic test for kinship or "race," which are socially constructed.

However, during the sixteenth to eighteenth centuries European people divided humans into categories, which they called "race," based on superficial, visual differences, especially skin, hair, and eye color. When, in the late nineteenth century, the idea of evolution began to be understood, it was combined with the assumption of human "races" into a theory that postulated that each "race" originated and evolved separately in isolated lines caused by geographic distance. White people believed that white people were superior to other "races" and even assigned African people to be the missing link to monkeys and apes, since no fossil missing link had been found.[19]

Scientists only began to understand how heredity really works by 1953, when James Watson and Francis Crick figured out the structure of DNA. (DNA—deoxyribonucleic acid—is a long chain of deoxyribonucleotides carrying genes whose molecular products make possible the replication of the chain.) Before this, the general public believed that hereditary material was carried in the blood. Many white U.S. soldiers in World War II thought they could have a black child if they received a blood transfusion from a black donor. This is why blood banks kept blood supplies separate until 1952. This idea is housed in our language, in words such as "full-blooded," "bloodlines," and "blood relatives." Since then scientists and social scientists have gradually come to their current understanding of the origin of genetic difference. In doing so, most have abandoned the term "race" as not biologically meaningful.[20]

The current explanation of genetic differences goes like this. The genetic mutation that created *Homo sapiens* originally happened in one person. It took generations to produce a small group of people with this gene or genes. As this population grew and spread, it seems to have experienced some bottleneck, or reduction to perhaps 15,000 people at about 70,000 years ago. This could explain our human uniformity.[21] The population recovered and expanded, fragmenting into separate populations, each with some genetic distinctiveness due to genetic drift or random gene mutations. This distinctiveness may have begun in Africa before the original diaspora of *Homo sapiens* into the world, or it may have begun later—this is currently unknown.

When these possibly distinctive groups spread out to inhabit the world and adapt to wide-ranging climates and ecological niches, genetic differences began to occur by natural selection as adaptations to local conditions. Examples of this are the large chests of Andean Indians, which help extract oxygen from thin air at high altitudes, and the compactness of Eskimos, which conserves heat.

Whether natural selection has shaped differences in skin and eye color and in hair is more difficult to answer. For example, no single gene has yet been located for skin color; it seems to be the result of several related genes. Skin color is determined by the amount of a pigment called melanin in the skin. Genes determine the amount of melanin produced. Some biologists argue that, since more melanin protects the skin from sunburn and skin cancer, people with genes for producing more melanin do well in sunny climates. Melanin, however, also slows the production of vitamin D that occurs when skin is exposed to sunlight. When people with dark skin—i.e., much melanin—move into climates with less sunlight, they are at risk for vitamin D deficiency. They may also be at greater risk for frostbite. Somehow natural selection in areas away from the equator favors those with genes for producing less melanin, and lighter-skinned people over time have come to predominate.[22]

Others argue that protection against sunburn and skin cancer has a trivial effect on reproductive success and point out that there are at least eight current theories about why people from the tropics have dark skins. Possibly it is sexual selection, in combination with natural selection, which produces much of our visible variation, just as Darwin argued.[23]

Sexual selection would work by selecting traits that have no direct survival value but that indirectly enhance survival by attracting a mate. Darwin argued that people pay much attention to breasts, hair, eyes, and skin color in selecting their mates, and they choose what is familiar to them. Current studies seem to confirm this.[24]

We may conclude that genetic differences developed over time, both by natural and by sexual selection, in groups of people isolated from each other. Groups based on external signs of genetic difference may have been on the way to becoming separate species. Since the species of *Homo sapiens* appeared, however, no new species has arisen within it. We are still a young species, unlike birds or ducks or chimpanzees, which have had the time and the isolation to develop multiple species. Every person alive belongs to one species, *Homo sapiens*. With lessened geographic isolation in modern times, genes are flowing all over the place, and many people seem ready to stop placing value judgments on skin color.

Unanswered Questions

1. How are we to evaluate life in the hunting-gathering period?

How we evaluate the hunter-gatherer life is an important judgment, for it reveals many underlying values. Anthropologists as a group have in the last thirty years radically changed their opinion of the hunting-gathering life. Until the 1960s, anthropologists routinely referred to its mere subsistence economy, its incessant quest for food, and its limited leisure save in exceptional circumstances. This dismal view may go back to the perspective of agricultural people, who found their way of life greatly superior. It also derives from the modern bourgeois perspective that focuses on the lack of material goods in the hunter-gatherer life.

In 1972 Marshall Sahlins published a book that offered a transformative point of view. Sahlins viewed hunting and gathering positively, calling it "the original affluent society" and analyzing how, in fact, people usually obtained their necessities without working more than five to six hours a day. Sahlins countered the traditional view that saw human history as progress from forager to farmer to industrial worker.[25]

A revolution in biology in the 1960s added to the reevaluation of the hunting-gathering period. Biologists began to claim that individuals act to benefit their offspring and the continuation of their genes rather than to benefit their group, their family, or themselves. Most biologists now hold that individuals are programmed to preserve their genes, that culture is a "canalized expression of human instincts."[26]

2. How fast does evolution occur?

Many genetic changes are neither beneficial nor disadvantageous, but are selectively neutral. Other genetic changes are advantageous or disadvantageous but only modestly; a beneficial gene may take thousands or tens of thousands of generations to be substituted. (In humans, 1,000 generations would span about 25,000 years.) Yet a genetic change may present such a strong selective advantage that it can be spread by natural selection in only a few thousand years.

One example in humans is the development of lactose tolerance, or the ability to digest lactose, the sugar found in milk. The majority of humans lose this capability at about four years of age, but lactose tolerance into adulthood has evolved in a minority of people, including the Tutsi of Rwanda, the Fulani of West Africa, the Sindhi of north India, the Tuareg of West Africa, the Beja of eastern North Africa, and some European groups. Since these populations herd sheep, goats, or cattle, there is a strong selective advantage to being able to digest lactose into adulthood. Animals have been domesticated only in the last 10,000 years (400 generations); within that time lactose tolerance has reached 80 to 98 percent in herding populations in which the adults drink milk.[27]

3. How much gene flow (mating among disparate groups) has there been?

Genetic research is beginning to reveal the history of our ancestry. Bryan Sykes, in *The Seven Daughters of Eve*, describes seven maternal clans that include more than 95 percent of native Europeans, based on tracing mitochondrial DNA. Mitochondrial DNA comes only from the mother. Mitochondria are tiny structures that exist within every cell but not in the cell's nucleus; they're found outside the nucleus in the cell's cytoplasm. Their job is to help the cell use oxygen to produce energy. The cytoplasm of a human egg is stuffed with a quarter of a million mitochondria, while the sperm has very few, which it jettisons once it enters the egg. Hence, all human beings receive their mitochondria from their mothers, and mothers make this extra investment in their children from the onset.

The genetic tracing revealed that the genes of almost 5 percent of native Europeans remain unaccounted for, people whose deep maternal lineage tells a different story. A primary school teacher in Edinburgh, Scotland, for example, carries the unmistakable signature of a Polynesian mitochondrial DNA. She said she knew her family history for 200 years and had no clue how this happened. But perhaps she was the descendant of a Tahitian woman who fell in love with a ship's captain, or of a slave captured by Arabs on the coast of Madagascar. A dairy farmer in Somerset, England, carried unmistakable African DNA, possibly a legacy of Roman slaves from nearby Bath. These examples illustrate gene flow, the mixing of genes by people traveling long distances. It used to be thought that this happened only recently with improved transportation, but now it is clear that people have been traveling and genes have been flowing deep into the history of our species.[28]

PART II

Ten Thousand Warm Years

5

Early Agriculture

(8000–3500 BCE)

Strikingly, agriculture arose independently in at least four and possibly seven places around the world in a time frame of about 8,000 years. Before 10,000 years ago (8000 BCE) virtually everybody lived on wild foods. By 2,000 years ago, the overwhelming majority lived by farming. Viewed against the 5 million years of hominid history, or even the 100,000 to 200,000 years of *Homo sapiens*, these 8,000 years represent an amazingly rapid rate of change, so rapid that historians call it the agricultural revolution, a fateful transition in human history.[1]

Why would people successfully adapted to a hunter-gatherer style of life abandon it in favor of farming all over the world within a few thousand years? This complex question brings us ever closer to our own time, for the turn toward agriculture began only about 400 generations (10,000 years) ago.

We are so accustomed to enjoying the food produced by farming that it is difficult to imagine ourselves sitting around the fires of hunter-gatherers. Unpalatable though their diet may seem to us, archaeologists now believe that changing to domesticated food may have represented a decline in the quality of diet, and certainly it meant an increase in the per capita workload.

Why would people do it? No one knows for certain, but much new evidence has come to light in the last thirty years. The simple answer seems to be that people had to, in order to survive. Those who did not make the switch died.

The complex reasons for why people would gradually abandon hunting and foraging can be discussed around the idea of a food crisis. The first pri-

ority for humans, like other animals, is to locate enough food; the more
they find, the more children they have, keeping them ever urgently seeking
food.[2]

By about 9000 BCE groups of people in many places began to feel a
pinch in their food supplies. There were no longer as many abundant areas
into which to move; other people were already there. From an estimated
50,000 *Homo sapiens* at the time of the expansion from Africa, the human
population had reached 5 to 6 million by 9000 BCE. During the period of
hunting and gathering, the population had grown slowly, but over time the
increase was significant. The Earth may have been reaching its human carrying capacity under the techniques of the hunting and gathering life.

Population pressures were not the whole story, however. As described
in the last chapter, the climate on Earth was changing faster than usual.
From about 9000 BCE the last ice age receded as temperatures rapidly
warmed around the globe. This warming affected people in many ways;
humans everywhere used their ingenuity to take advantage of the new
circumstances as rising seas pushed people inland and rising temperatures
changed plants and animals. Building on their previous achievements—
the use of fire for cooking and clearing land, the use of language for
social cooperation, the development of tools for solving problems—and responding to what plants and animals did, many people over the next few
thousand years transformed themselves, and were transformed, from
wandering bands of hunter-gatherers into settled villages of herders and
farmers, who could produce a surplus of food, at least temporarily.

Plants and Animals Enter Domestication

Humans are not the world's first, or only, farmers. Ants grow plants (fungus) and tend animals (aphids). They collect seeds and store them in cham-
bers near their nests. At least 225 genera of plants are dependent solely on
the activities of ants for their propagation. Like ants, people became involved in the life cycles of certain plants and animals, and what we call agriculture resulted.[3]

In the past fifty years the work of archaeologists has greatly increased
our knowledge of the beginnings of agriculture. For evidence they rely on
the remains of animals (bones) and plants (seeds and pollen). For example,
the warming of the climate 11,000 years ago was documented in 1968 by
the analysis of pollen in two lake beds in Iran. Other evidence is often fossilized human feces, called coprolites, which reveal what plants have been

eaten. The best sites for recovering this evidence are located in dry regions. Radiocarbon dates are reliable to within a few hundred years if different materials from the same site yield the same dates.

The domestication of animals and plants was a long, unintended, reciprocal, evolutionary process. Various local groups of people may have exchanged ideas about how to do it, but it seems to have occurred independently at different times in at least four different locations: southwest Asia (the Fertile Crescent), China, and southeast Asia; Africa; and the Americas (Fig. 5.1).

The reason for the simultaneous appearance of agriculture in different areas was the warming climate, during which the plants and animals that survive most frequently are those that display traits of flexibility and nonspecialization. These traits characterize the young of most species; hence, warming climate produces animals that retain their juvenile traits (called neoteny), including docility, lack of fear, dependence, and early sexual maturity. Animals evolved the traits that predisposed them to domestication.[4]

Domestication can be defined as a kind of genetic engineering in which humans gradually take control of the reproduction of an animal or plant that is predisposed to engagement with humans, separating it from its wild species in order to control its development into a new species with the characteristics that humans desire.

The first animal to enter into domestication was none other than people's best friend, the dog. The ancestors of dogs are gray wolves, found

5.1 The Origins of Agriculture
(*Source:* David Christian, 2004, *Maps of Time: An Introduction to Big History*, Berkeley, CA: University of California Press, 213.)

around the world after evolving in North America. Wolves gradually evolved into dogs in the Americas in about 11,000 to 10,000 BCE and in present-day Iran a little later. It is easy to imagine that dogs, as the climate changed, also hung around human campfires and hunting sites looking for food and interacting with people. Dogs adapted themselves easily to human activities. They were pack animals that followed a leader, and accepted a person as a surrogate pack leader. Captured puppies could easily be cared for until adulthood. As tame adults, dogs helped in hunting and later, as other animals were domesticated, played a vital role as guards against predators and as allies in herding. Dogs were helpful scavengers, cleaning villages by eating human feces. They were eaten in some cultures but not in others.

As one might expect, the domestication of cats occurred much later, even though cats evolved into their present state as long as 3.4 to 5.3 million years ago. It was probably Egyptians who domesticated cats in order to protect their granaries against rodents, a practice documented by 1500 BCE. Although cats are solitary as adults, they are sociable as juveniles; that seems the key to their domestication. Domestic cats have been documented in Greece and China from 500 BCE.

Only about thirteen large mammals (those weighing over one hundred pounds) permitted themselves to be domesticated. The big five were sheep, goats, cows, pigs, and horses. The other eight were two kinds of camels, donkeys, llamas, reindeer, water buffalo, yaks, and Bali cattle. All these animals entered into domestication between 8000 and 6000 BCE. All shared the following characteristics: they ate plants, grew quickly, bred in captivity, would not kill their keepers or themselves trying to escape, and had a social structure (herd) that made them easy to manage. Most large mammals were not willing or genetically suitable for domestication; otherwise we might have hippopotami producing our milk and be riding zebras in our parades.

Each region in which herding emerged had its distinctive animals. The earliest region was southwest Asia—or the Middle East, as it is often known in the United States.*

---

* Terminology regarding this area is not consistent. The Near East is used to refer to areas bordering, or nearly so, the eastern end of the Mediterranean Sea. This would include the modern territories of Turkey, Cypress, Syria, Lebanon, Palestine, Israel, Jordan, and Egypt. The Middle East can refer solely to countries bordering the Persian Gulf: today's Iraq, Iran, Kuwait, Saudi Arabia, and other Gulf states on the Arabian Peninsula. But news commentators today use Middle East to include the Near East as well as the Gulf states. A term also widely used by historians is the Fertile Crescent, referring to an arc of land that runs northward from parts of modern Israel, Jordan, and Lebanon, curves eastward along the border of Turkey and Syria, then turns south along

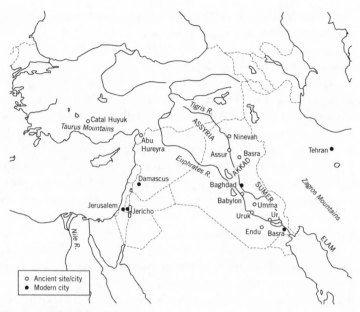

5.2   Ancient Southwest Asia

In certain areas of the Fertile Crescent people were able to settle in villages without the domestication of plants or animals. They collected and stored enough wild grains to supplement their abundant hunting, mainly of gazelles. This step is called complex foraging, as contrasted with simple foraging, which is characterized by no storage or long-term settlement.

Two wild animals in the Fertile Crescent were willing to get closely involved with humans—sheep, beginning in about 9000 BCE, and goats, in about 8000 BCE. Since both sheep and goats were able to digest many more kinds of grasses and foliage than humans, they became an efficient way for turning inedible plants into protein for people, who gradually learned to herd them from area to area and eventually to enclose them and protect them from predators. The willingness of sheep and goats to become dependent on humans ensured their evolutionary success. The population of each now exceeds 1 billion, while their wild counterparts teeter on the brink of extinction.

---

the Zagros Mountains on the borders of Iran and Iraq. I use Middle East in its broad, inclusive sense, or the Fertile Crescent, as defined here. I use Near East to specify just the eastern end of the Mediterranean, and Mesopotamia to refer to just the Tigris and Euphrates river valleys (Fig. 5.2).

The process of domesticating sheep and goats may have started with men guarding a herd as it moved. Men then herded groups of animals within certain locations, began feeding them, and then corralled and housed them in permanent settlements.

The domestication of plants took place in an equally long, slow process. People carefully observed the wild grasses as they collected seeds to grind and eat. According to the rubbish heap theory of plant domestication, they first noticed wild seeds growing at their campsites where the seeds not consumed had been dumped. Women probably carried out the early sequences in plant domestication, since they were usually the gatherers in hunter-gatherer groups.[5] They must have observed that some grasses had larger seeds, easier than others to harvest and process into foods. Some grasses had heads that shattered easily and dispersed their seeds; others held the seed firmly until maturation.

Women in the Fertile Crescent learned to seek out three kinds of wild grasses—emmer wheat, einkorn wheat, and barley—as well as two forms of wild legumes, lentils and chickpeas. Gradually, after gathering these wild forms, women learned to tend and protect them. They noticed where they grew and where the seeds sprang up the next year. Eventually women learned to save some seeds and sow them in areas in which they would grow, to water and weed them, to choose the largest seeds and the healthiest plants, and to store the surplus. Men continued to hunt, and women supplemented their kill with increasing supplies of wheat, barley, and peas.

By about 7500 BCE in the Fertile Crescent, people settled in permanent villages, tending crops and animals. Keeping goats and sheep and growing wheat and barley, they were able to produce more food from a smaller area than they had been able to by hunting and gathering. In this gradual process, many villages of complex foragers were abandoned as people returned to simple foraging; not all villages could or would make the transition to agriculture. There is evidence of extensive use of female infanticide as early villagers struggled to limit their population to the food available.

In statistical terms, one hunter-gatherer needed about ten square miles of favorable territory to collect enough food to live. One square mile of cultivated land, however, could support at least fifty people. Hence agriculture could support a human density fifty to a hundred times greater than hunting and gathering could.[6]

By about 6000 BCE settled life was becoming the norm in the Fertile Crescent. All the suitable crops and animals in the region had been domesticated, and they became the basis for the adoption of agriculture in adjacent areas: Europe, where different adaptations were necessary, and in the Nile Valley, almost unchanged.

In 6000–5000 BCE Greece and the southern Balkans, where the climate was similar to the Near East, shifted to agriculture and probably domesticated cattle. Archaeologists have hotly debated whether agriculture spread by people circulating the knowledge by word of mouth or by people themselves moving into new areas. But genetic research has revealed unequivocally that people themselves moved rather than just talked about how to do it.

The movement of agriculture into central and northwest Europe took about 3,000 years after its adoption in Greece. By 4000 BCE farming had moved into the river valleys of central Europe—the Rhine/Danube and the Vistula/Dniester areas. Between 3000 and 2000 BCE it had been adopted in northwest Europe and a thousand years later in Denmark and southern Sweden. In these areas forests had to be cleared by slash and burn techniques, with permanent fields coming later when population pressures increased. Oats and rye, which grew as weeds in the Middle East, proved to be the food crops that flourished in the cooler, wetter climate of northwest Europe.

As farmers fanned out from the Middle East and Turkey, they took with them their language, called Indo-European. One of perhaps ten proto-languages spoken in the world at that time, Indo-European was used in parts of the Near East and around the Caspian and Black seas from about 8000 to 2000 BCE. Sanskrit evolved from it in about 1500 BCE or earlier, as did Greek in about 1450 BCE.

Agriculture waited some 2,000 years after its start in the Middle East to begin in the Nile River valley, in about 4300 BCE, based on barley, wheat, and cattle. Why farming took so long to get under way in a valley climatically suited for it is a puzzle. Cattle probably were domesticated independently in the Sahara as early as 7000 BCE, but as it dried up after 6000 BCE, cattle herders were forced to the fringes.

Africans domesticated the ass, as a beast of burden; the guinea fowl, a favorite dish of ancient Egypt and later of Rome; and cats, as mentioned earlier. Millet, sorghum, wild rice, yams, and palm oil are other foods domesticated in Africa. Yams belong to those plants that are propagated not by seeds but by stem cuttings, tubers, or roots. These plants include maniocs, bananas, sugarcane, and taro. Since these crops leave no seeds as evidence to find, African and Asian people may have cultivated them much earlier than can be known.

In Asia the evidence for early food production remains sketchier, possibly because the climate is warmer and wetter than in the Near East. The accepted picture is that millet and rice were domesticated in China in about 6000 BCE; soybeans did not appear until about 1100 BCE. Pigs and poultry were domesticated there. Rice seems to have been domesticated independently in India and possibly also in Southeast Asia.

People in the Americas developed their own gardens. By 6000 BCE people in the highlands of Mexico were cultivating up to thirty plants for food, medicine, and containers. These plants included maize (corn), chile peppers, tomatoes, five kinds of squash, gourds, avocadoes, papaya, guava, and beans. Maize emerged slowly; gene study shows that domestication began about 7000 BCE. In the wild the cob was about the size of a human thumb. Gradually larger cobs with higher yields were developed, until about 2000 BCE when the production of maize sufficed to support village life. Since there were no suitable animals to be domesticated, other than dogs and turkeys, hunting continued as long as possible. Cotton and peanuts were also cultivated.

In the mountains of Peru (including large parts of present-day Bolivia and Ecuador) another set of domesticated crops and animals developed. The llama and alpaca were used as beasts of burden, not as food. People based their diet on potatoes and quinoa, a protein-rich seed grain. Maize spread to Peru by about 1000 BCE.

In the long view of time, the domestication of plants and animals leading to agriculture as a mode of production occurred nearly simultaneously in various parts of the world. In the short view of time, however, within a few thousand years some areas lagged behind others with fateful consequences. Because people in the Americas had no suitable grains and animals for early domestication, the evolution of complex societies there began 3,000 to 4,000 years later than in the Middle East, Europe, and Asia. As a consequence, when Europeans arrived in the Americas in 1500 CE, they found societies in many ways comparable to those of the Middle East in about 2000 BCE. With their horses, guns, and diseases, products of their more evolved agrarian societies, Europeans were able to strangle the more slowly emerging civilizations of the Americas.[7]

People's experiments with plants between 9000 to 3000 BCE were so successful that no new basic food plants have been domesticated since then. The only exceptions seem to be cranberries, blueberries, and pecans, which were gathered by native North Americans but have been domesticated only in the last two centuries.

Out of approximately 200,000 species of flowering plants, only about

3,000 have been used extensively for human food. Of these, only fifteen have been and continue to be of major importance: four grasses (wheat, rice, maize, and sugar), six legumes (lentils, peas, vetches, beans, soybeans, and peanuts), and five starches (potatoes, sweet potatoes, yams, maniocs, and bananas).[8]

Three Small Towns

One excavation in particular reveals the process of people turning to farming as a way of life: the village of Abu Hureyra in modern-day Syria. This site was first occupied in about 11,500 BCE as a small village settlement of pit dwellings with reed roofs supported by wooden uprights. Inhabitants collected and stored wild barley, wheat, and rye. They hunted Persian gazelles that arrived from the south each spring, killed them en masse, and stored their meat, preserving it by drying and salting. Abu Hureyra slowly increased to 300 or 400 people who, in about 10,000 BCE, when the climate temporarily cooled, abandoned their village to return to nomadic life—still an option when difficulties like cooling of the temperature or depletion of the adjacent firewood arose.

About 500 years later (about 9500 BCE) another village arose on the same location. At first the village inhabitants hunted gazelles intensively, but in around 9000 BCE they switched to herding domesticated sheep and goats and cultivating wheat, chickpeas, and other cereals. They built rectangular, one-story mud brick homes of more than one room, joined by lanes and courtyards. The homes had floors of black, burnished plaster, occasionally decorated with red designs. They seemed to be dwellings for a single family. The town, eventually covering nearly thirty acres, was abandoned by about 5000 BCE for no known reason.

Two other villages in the Middle East that became towns in these early years have been extensively excavated—Jericho on the west bank of the Jordan River and Çatal Hüyük (pronounced Cha-TAHL-hoo-YOOK) in central Turkey.

The settlement at Jericho, at a bubbling spring, extended over at least 9.8 acres by 7000 BCE. Here people built a massive wall around their settlement. They cut down into the rock about 9 feet (2.7 meters) deep and about 10 feet (3.2 meters) wide, then bordered it with a stone wall 10 feet high, complete with 25-foot towers. Inside the wall were clustered beehive-shaped mud brick homes. Why people built such a wall remains a mystery; possibly it was a means of flood control or a defense against other people eager to steal food. The wall attests to communal labor well organized and supported by surplus food.

At Jericho the dead were deposited within the settlement, often with the head severed from the rest of the body. Sometimes modeled heads were made in painted plaster. These may have indicated differences of status; there were no other indications in the burial goods.

Settlers at Jericho herded sheep and goats that, by about 6500 BCE, constituted 60 percent of all meat consumed. Presumably the gazelles were becoming depleted. Cattle and pigs seemed to be coming under increasing control by people, who were also cultivating wheat, barley, lentils, and peas, rotating crops to keep yields high. Considerable trade occurred—obsidian from Turkey, turquoise from Sinai, and seashells from the Mediterranean and Red seas. Small clay spheres, cones, and disks suggest a simple recording system for keeping track of something—perhaps commodities traded?

The obsidian used at Jericho came from Turkey, probably from its largest trading center, Çatal Hüyük. Some of Çatal Hüyük's prosperity resulted from the trading of obsidian quarried in nearby mountains. Obsidian forms when molten lava flows into a lake or ocean and cools rapidly, producing a glassy rock. It was much valued for its propensity to split, creating sharp edges, and its ability to be polished brightly in the creation of high quality tools, weapons, mirrors, and ornaments. The site at Çatal Hüyük, first excavated in 1961 to 1963 CE, revealed the details of how humans adapted to a settled existence and created beauty in doing so.[9]

Çatal Hüyük spread over an area of thirty-two acres on the Konya Plain of south central Turkey near a marsh surrounded by well-wooded areas. Excavations show that it was rebuilt, presumably when houses began to crumble, at least twelve times between 7000 BCE and its abandonment around 4500 BCE. Constructed of sun-dried mud bricks made in molds, the houses were designed to back into each other with occasional courtyards. The roofs were flat and were entered by a ladder to an opening in the roof. The outside walls of the outermost houses provided a kind of defense for the town.

The food supply in Çatal Hüyük was based on domesticated sheep, goats, and pigs, and on two kinds of wheat, barley, and peas. Some hunting of red deer, boar, and onager went on, while some wild plant life, such as grasses and acorns, was collected and stored. There are some indications, not conclusive, that flax (a plant from which linen and linseed oil is made) may have been grown.

Men in Çatal Hüyük grew to an average height of 5'7", while women reached an average of 5'2". Men lived an average of 34 years, while women averaged 30 years. These averages include a high rate of infant and child death. The skeletons found at Çatal Hüyük reveal some arthritis but no rickets or vitamin deficiency. An overgrowth of the spongy marrow space of the skull does reveal, however, that about 40 percent of the adults studied suffered from anemia, which implies that malaria was endemic. The population is estimated as having started at about fifty people in 6500 BCE and reaching nearly 6,000 in about 5800 BCE, but this estimate is shaky.

The objects found in Çatal Hüyük suggest a high level of achievement in creative activities. The town, too small in population to feature the specialization found later in cities, seemed to have no dominant class or centralized political structure. Burial goods indicate social equality. Apparently everyone participated in the creative, artistic activities made possible by a settled life still focused on hunting and gathering but supplemented by herding, cultivating, and trading.

People in Çatal Hüyük made coil-based pottery, not yet the wheel-turned variety. People constructed baskets and wove textiles of wool or flax. They chipped exquisite knives and spears, carved stones and bones, worked leather and wood, and created jewelry and cosmetics. Objects of copper and lead, which occur naturally in almost pure form, were found at Çatal Hüyük as decorations and ceremonial objects.

The representational art at Çatal Hüyük reveals how powerfully people were still oriented toward hunting. Paintings on plastered walls depict hunting scenes, with men and women draped in leopard skins. Other scenes depict vultures cleaning bones, apparently human ones. Men were buried with weapons rather than with farm tools.

Women were often buried in special rooms that archaeologists interpret as shrines. Forty of these rooms have been excavated—one for every two houses in Çatal Hüyük. They feature sculpted heads of wild bulls, relief models of bulls and rams, depictions of female breasts, goddesses, leopards, and handprints. Since fat, fertile, female figures far outnumber those of males, experts believe that inhabitants gave top honors to a goddess. Since women are buried in these rooms, one may deduce that they created the rituals and served as priestesses. Some pictures depict a woman giving birth to a bull. Several sculptures show a woman with each arm resting on a leopard, apparently with the head of an infant appearing between her legs. Do the leopards, representing death, depict death intertwined with life? Or does the leopards' support show the power of the goddess over nature? It's anyone's guess.

What people in Çatal Hüyük believed about death cannot be known, except that food offerings have been found with the bones, suggesting they believed in an afterlife. Their murals indicate that after death, the bodies of people in Çatal Hüyük were exposed to vultures. When the bones had been cleaned, they were buried in the shrines or under the sleeping platforms in the houses that they had occupied in life.

For unknown reasons people abandoned the site at Çatal Hüyük sometime in the fifth millennium BCE.

Effects of Settling Down

As people began to settle in villages and towns to cultivate plants and tend to animals, unforeseen transformations occurred in their lives. The complexity of these changes, which we live with still, defies analysis and can only be suggested by describing certain aspects of social life and some impacts on Earth itself.

The advantage of agriculture—more food per unit of land—meant that people had to figure out how to store and preserve food. They had to defend their towns against large animals and other people, since they had too much to lose to pull up and move on. They needed people who specialized in creating storage (pottery, baskets, and storage bins) and in defending the town. Surplus food could be used to support these specialized people. It could also feed babies; with cereal available, they could be weaned earlier, and women could produce more children closer together.

Cultivation, however, also meant that people had to work harder. They had to learn to exercise internal restraints on themselves, such as working long hours when they would rather be sleeping or socializing, or not eating the best seeds on long winter nights because the seeds had to be saved for spring planting. People had to spend long hours grinding seeds and weaving cloth, possibly not their favorite activities. Once they had domesticated animals and plants, they, too, had become domesticated in a mutual exchange.

Once settled in towns, people had to interact with several hundred to several thousand other people, rather than with thirty to fifty. Modern analysis suggests that about 150 people is about the limit that an individual can deal with on a personal basis; after that rules, guidelines, and policies need to be developed. In larger groups people have to create mechanisms for solving disputes; protolawyers and protojudges emerge. Ceremonies are created, and the people who carry out the ceremonies assume a kind of authority. At some point the idea of private property arises—a house belongs to someone, or a patch of land, or a number of animals. New rules had to be made about ownership and control of land. Families had to be defined more tightly and decisions made about who would live with whom. People, no longer limited by what they could carry around, began to acquire more physical objects. Garbage and human waste began to be issues.[10]

Because textiles are perishable, they are not as well documented as pottery. The earliest known textile was found in 1993 in southern Turkey, at a place known today as Çayönü. It is a fragment of white cloth, about an inch and a half by three inches, wrapped around a handle of a tool made from antler. The cloth, semifossilized from contact with calcium in the antler, probably is a piece of linen, woven from the fibers of the flax plant. It has been dated at 7000 BCE by radiocarbon testing.[11]

Experts believe that from the time that people settled in villages they extended their basket-weaving techniques to crude fabrics. This consumed many hours of labor, possibly as many as pottery and food production combined. Cloth became an essential part of human society as clothing and adornment that indicated social rank.

As people settled down and reduced the portion of wild meat in their diet, they had to look for supplies of salt. The body of an adult contains three or four saltshakers worth of salt. The body loses salt through perspiration and cannot manufacture it but must replace it to live. Eating solely wild meat provides enough replacement salt, but adding cultivated crops to the diet meant that people had insufficient salt and had to locate it in the Earth. The animals they cared for also needed salt; a cow requires ten times the amount that a person does. Villagers could locate supplies of salt, probably by following wild animal trails, but people in towns, and eventually in cities, faced a bigger problem. Eventually salt became one of the first international commodities of trade and the first state monopoly—in China in 221 BCE.[12]

In the early days of plant and animal domestication, people saw that the females of every species produced new life. What did they think about the role of the male in producing offspring?

That humans were aware of men's role in reproduction cannot be documented until written history. Astonishingly, a few present-day hunting-and-gathering people seem not aware that men are required for human reproduction. Yet archaeologists believe that by herding animals, most groups had made sufficient observations to understand the connection.[13]

As people became involved in the daily lives of animals, they exposed themselves to animal diseases. Plant pests cannot make the leap to humans, but many animal diseases can. One form of tuberculosis came with milk from cows and goats. Measles and smallpox came from cattle. A form of malaria probably came from birds, while influenza came from pigs and ducks. These diseases have played a vital role in the human story.

Life in villages and towns became precarious in ways different from the life of hunting and gathering. The possibility of disease became a cause for anxiety. The vagaries of the weather became a daily concern: Would the rain fall at the right time? Would the temperature be too hot or too cool for the crops? Hailstones could flatten a crop, as could an attack of insects or fungi. Wild game could decline or disappear; a sudden flood could surge through. Human lives were always at stake.

People settling down must have felt at the mercy of the Earth itself. Their focus shifted from placating wild animals to honoring the source of life itself and asking assistance from this source. Associated with the shift from hunting to cultivation, archaeologists have found many figures of fertile females with huge breasts and buttocks. These figures have been found across the Middle East and central Europe, dating from the period of transition, from about 8000 to 3500 BCE.

It is impossible to know exactly what these artifacts meant to the people who created them, as discussed in chapter 4, but it is difficult not to conclude that some respect for fertility is being shown. Some people must have imagined the whole Earth itself as a goddess of fertility; this idea has come down to the time of written records as the Greek goddess, Gaia. Since women give birth to new life, the spirits that people appealed to seem to have taken the form of goddesses of fertility, so vital to newly settled people. Goddess figures are associated with early agricultural societies from the Aegean Sea to Indonesia, in the form of the rice goddess, Dewi Sri, daughter of Vishnu.

As cultivation began to produce a surplus of food, a few people could specialize in creating rituals and the art associated with them. As at Çatal Hüyük, these first specialists, or priestesses, seem likely to have been women.

Women scholars have conducted research over the past forty years hoping to find evidence of some societies in which women controlled political power. They have not found evidence for any such matriarchies. Apparently by the time human density had increased enough to require centralized political power, women were limited by the increasing numbers of children, and men were powerful enough as farmers, military leaders, and priests to control political power.[14]

And the number of children did increase. At the beginning of experiments in settling down, the world's total population was estimated at 6 to 10 million, about half the current size of Mexico City. By about 4000 BCE the rate of change had increased dramatically. By 1000 BCE the population was

approaching 50 to 100 million. Humans were off and running on a stagger-
ing adventure into density and complexity.

Some of the costs of switching to a settled way of life may have been apparent to people undergoing the changes. They were working harder and
were more at the mercy of the weather. They had more disease and less variety in their days. They may have told nostalgic stories about the hunting-
and-gathering days of their ancestors.

Other costs, however, must have remained hidden to them, apparent
only much later in the perspective of time. These were the costs of damage
to the environment, to the fertility of the Earth itself. Among these environmental costs was deforestation, which began occurring even before plant
and animal domestication, when people set fire to areas of trees to create
meadows that would attract grazing animals. People increased this burning of trees when they wanted to create areas for cultivation; until metal
axes there was no reasonable alternative. Trees were also burned for cooking and heating. Negligible at first, this destruction increased as the population slowly mounted.[15]

Destruction of trees also occurred through the intensified herding that
took place during periods of drought, when goats would climb trees to eat
foliage and consume every seedling. Where goats are herded regularly,
forests cannot regenerate. Sheep, too, are a force for destruction, since they
eat grass by the roots, pulverizing the soil.

The simplest kind of cultivation eroded the soil. The moment the
structure of the soil was disturbed, even by a digging stick, it became subject to being blown or washed away. Such effects, again negligible at first,
grew with the human population into calamities, as soil became depleted
and rivers silted.

A perspective on the damage to Earth by cultivation was offered in
about 1872 by the Native American leader, Smohalla, who protested a pro-
posal to turn his people, the Plateau Indians of the Northwest, from hunting to cultivation: "You ask me to plow the ground. Shall I take a knife and
tear my mother's breast?"[16]

Persistent Hunter-Gatherers and Nomads

Not everyone, as the words of Smohalla indicate, turned to a settled way of
life between 8000 and 3000 BCE. Many areas of the world were simply not
suitable for cultivation; the soil was too hard and unfertile, suitable grain
grasses were not present, and the rainfall and temperature would not per-

mit crops to mature to harvest. Other areas produced such abundance that farming was not necessary. In these regions of the world people continued their hunting and gathering, or they combined it with nomadic herding of some kind.

In the cold tundra regions of Eurasia it appears that reindeer were domesticated to pull people's sledges by 9000 to 7000 BCE. In the upper reaches of the Nile Valley and across the rift valleys and plains of eastern and southern Africa, people developed cattle herding cultures.

A key animal in herding turned out to be the horse. North of the warm fertile agricultural areas stretched vast grasslands, from central Europe to eastern Asia. This kind of area proved too cool for agriculture. People in grasslands continued their hunting and gathering until all the large mammals died out except the horse, which did most of its evolving in North America, then went extinct there. In about 4000 to 3500 BCE the people of the grasslands, or steppes, of southern Ukraine began to protect and feed horses in exchange for milk for human babies, dried dung to burn as fuel, and meat, especially in winter when food was scarce. Thanks to this relationship, the population of both humans and horses in the sparsely settled grasslands began to increase. Later, with the development of iron and the invention of the stirrup in central Asia in about 500 BCE, the horse nomads of central Asia would become a central force in human history, trading with and sometimes plundering settled areas.

The peoples of the grasslands of North America did not domesticate horses, which died out on that continent where they had originated, the victims of changing climate and hunting by humans. People there continued hunting and gathering with some cultivation of plants first domesticated in Mexico until horses were reintroduced by Europeans after 1500 CE. Many people in South America continued hunting and gathering, with cultivation in the Peruvian Andes and perhaps in some tropical areas.[17]

The decision of a group of people to commit itself to cultivation and settlement can never have been an easy one. Even if settlers succeeded in producing enough food for themselves, they still faced raids by herders or hunter-gatherers desperate for their stores of grain and animals. Intergroup conflicts became more frequent and threatening.

The oldest literature in the world contains stories of conflicts among hunter-gatherers, herders, and agriculturalists and the conflicts within individual people as they made these choices. The most ancient of this literature, *The Epic of Gilgamesh*, was written about 2100 BCE by the Sumerians, the people of the first city-state, Sumer, located at the mouth of the Euphrates River in present-day Iraq. The stories in oral form went back at least to the seventh millennium (6999–6000 BCE) when people were beginning the domestication of plants, animals, and themselves.

The Epic of Gilgamesh is organized around the life of a historical man, Gilgamesh, who ruled the city of Uruk sometime around 2750 BCE. The existence of the epic has been known to the modern world only in the last 130 years, since shortly after cuneiform writing was deciphered in 1857 from clay tablets found at Nineveh, the ancient capital of Assyria, north of Uruk in present-day Iraq. At present only about two-thirds of the text is available in continuous form; other sections contain many gaps.[18]

In the story, Gilgamesh is portrayed as a superhero, one-third human, two-thirds divine, extraordinarily handsome and strong. The goddess of creation, Aruru, has also created a wild man, Enkidu, strong and handsome as well who lives in nature, wears animal skins, and drinks water with the gazelles.

Gilgamesh sends Shambat, a sacred prostitute and priestess of the goddess Ishtar, to seduce Enkidu and bring him to the city. She teaches him to wear clothes, cut his hair, drink wine, and be civilized. Enkidu challenges Gilgamesh to a wrestling match in which Gilgamesh prevails.

After Enkidu has been assimilated into urban life, he and Gilgamesh go on adventures together. First they kill the demon, Humbaba the Terrible, who guards the Cedar Forest, and they cut down the sacred cedar trees. Back in the city, the goddess Ishtar wants Gilgamesh to marry her. When he refuses, she sends down the Bull of Heaven to wreak havoc in the city. The friends manage to slay the bull. Their relatives and their dreams warn them beforehand that they should not cut down the trees or kill the bull. The gods are not amused with these deeds and in punishment decree that one of the friends, Enkidu, must die a long and painful death, attended by Gilgamesh.

These stories in *The Epic of Gilgamesh* seem to express the ambivalence that people were feeling about cutting down forests and domesticating wild bulls. The same conflicts are echoed in the ancient Hebrew myth of Adam and Eve in the Garden of Eden. This story was not written down until about 1000 BCE, but it was based on stories that circulated much earlier in Babylonia, which followed Sumerian civilization at the mouth of the Tigris and Euphrates rivers in present-day Iraq.

In the story of Adam and Eve, people are presented in their original condition as hunters and gatherers in the natural world, the Garden of

Eden, where food can simply be gathered in abundance. The metaphor of
the garden refers not to a human-made garden but to God's natural garden
before agriculture.[19]

In the story, Eve presents Adam with an apple from a tree representing
the knowledge of good and of evil. God has specifically forbidden the fruit
to people. When Adam consents to eating the apple, God punishes the
couple by banishing them from the abundance of nature and condemning
them to laborious toil in the production of their food.

In this story about the transition from foraging to herding and farming, the apple tree represents plants once gathered but now cultivated, primarily by women. The apple tree may not only be symbolic but also
specific; apples seem to have originated in the mountains of Kazakhstan
and spread across to the Caucasus Mountains.[20]

The apple tree also represents the knowledge of good and evil in the
sense that when people settle down to a life of cultivation they must designate which of their behaviors are good (helpful) and which are evil (harm-
ful). They must develop rules and guidelines for human behavior in larger,
settled groups with a surplus of food for the first time. Previously, in smaller
groups, the whole range of human behavior could be mostly accepted, and
when it could not be, people could leave for other groups or start a new one.
Elaborate rules about behavior were not necessary.

In the story of Adam and Eve, God is not pleased with people learning to cultivate plants. He punishes them by throwing them out of his
garden and forcing them to work for their food. The narrator seems aware
of the costs of settling down—the increased labor and the constraints on
behavior.

The story continues with the sons of Adam and Eve, who are Cain and
Abel. Cain, the elder, becomes a tiller of the ground, while Abel chooses to
be a keeper of sheep. When they each offer a gift of their labor to God, he refuses the gift of Cain but accepts the animals of Abel's flock. Out of jealousy,
Cain kills Abel, becomes alienated from the ground, and is forced to return
to wandering. The God of the Hebrews is not pleased with the farmer, with
the transition to agriculture. The Hebrews remained a herding, nomadic
people until after they conquered the Canaanites, who were agriculturalists worshipping goddesses; eventually the Hebrews, too, had to settle
down into agriculture.

In the metaphors of *The Epic of Gilgamesh* and the story of Adam and
Eve, we hear the laments of people who are no longer hunters and gatherers, people conflicted about their new arrangements and haunted by
doubts about what they have done. Yet they press forward to create what

they can from the abundance of the soil, the water, the sun, and all the plants and creatures sharing the living Earth. By about 3500 BCE something we call civilization has emerged.

Unanswered Questions

1. Did the Americas and Asia remain in contact after the initial colonizing of people from Asia?

This tantalizing question remains open, for there is some evidence, but not sufficient to be conclusive, that contact continued. The Chinese melanotic chicken, with black bones and dark meat, existed in the Americas, where it was sacrificed but not eaten, just as it was in China. A few scholars argue that the Mayan calendar probably came from Taxila in present-day Pakistan, and that four of the twenty names given to the days of that calendar were borrowed from Hindu gods. Some evidence suggests that the peanut plant found its way from America to coastal China and that American cotton got to India. Perhaps more evidence on this question will be forthcoming in the next decades.

2. Can the Garden of Eden and the Biblical flood be located historically?

Scholars believe that the Garden of Eden may have been located along the shores of Iraq where the Tigris and the Euphrates rivers flow into the Persian Gulf. This shoreline must have provided abundant food to hunters and gatherers, but as the seas rose people were forced to move inland to drier, already populated areas. Some may have found refuge on the edges of the small lake that would become the Black Sea. Melting water from the glacier did not flow down the Dniester, the Dnieper, the Don, or the Volga rivers because it was being diverted westward by a bulge in the Earth's soft interior formed by the previous weight of the glacier. In about 5600 BCE the rising water of the Mediterranean Sea suddenly overflowed to the northeast, forming the Black Sea over the course of two years, forcing refugees to move in every direction, including into present-day Iraq. Whether this constitutes the Biblical flood is pure speculation, but the timing is right for the story of that flood to circulate orally until recorded in the Old Testament documents.[21]

6

Early Cities

(3500–800 BCE)

O nce people had developed food production to the point of being able to store surpluses, the human population began to grow more rapidly; from 8000 to 3000 BCE it increased from 6 to 50 million. Some people began to live in cities of 10,000 to 50,000 inhabitants. In cities, people created a whole new range of ideas and structures that came to be called "civilization" (from the Latin *civitas*, for city) by Western historians. Among the characteristics of "civilization" usually cited are: storage of food, development of a priestly caste, central authority, nonagricultural specialists, social stratification, increased trade, development of writing, tribute forcibly collected from outlying farmers, development of soldiers and standing armies, monumental public works, and increased gender inequality.[1]

Historians have widely debated the possible meanings of civilization, and currently many avoid the term, especially since, during recent colonial times, historians in imperial countries used "uncivilized" to designate colonial people. In addition, attitudes toward civilization are changing. Once the rise of civilization was thought to demonstrate humanity's success in overcoming its savage nature. Now, many are beginning to wonder if civilized life is as savage, if not more so, than the hunting-and-gathering life, particularly in its social inequality and frequent warfare.

Some historians replace civilization with the term "complex society." I mostly use "cities" or "urban life" to indicate the early complex societies that arose across Eurasia at about the same time; whenever I use "civilization" or "complex society," I am referring to the constellation of characteristics of urban life as listed earlier, without implying a value judgment either negative or positive. Also, I like the suggestion of David Christian that we call these first states "agrarian civilizations" to remind ourselves that they depended on their rural hinterlands for food and tribute.[2]

The first cities arose more or less simultaneously in about 3500 BCE in river valleys in four areas of Afro-Eurasia, the continents of the earliest human habitation. These earliest cities appeared in the Tigris and Euphrates valleys in southern Iraq, in the Nile valley in Egypt, in the Indus valley in Pakistan/India, and, a bit later, in the Yellow (Huang) River valley in China. Urban areas appeared later in the Americas, beginning with the Olmecs in Mexico in about 1300 BCE and with groups in the Andes in about 900 BCE. The fact that cities arose later in the Americas than in Afro-Eurasia had significant consequences, to be discussed in chapter 10. In this chapter I will describe the first cities in the Tigris/Euphrates valley in some detail, then generalize about life in the other three Afro-Eurasian urban areas.[3]

The Sumerians

As described in the previous chapter, the earliest domestication of wheat, barley, sheep, and goats took place in the highlands of Turkey, Iraq, and Syria; later these crops were taken down to the fertile Tigris and Euphrates river valleys, where irrigation of wheat and barley proved necessary during the dry months. Sometime before 5000 BCE people in the river valleys figured out reliable systems of irrigation.

Out of these agricultural people there emerged, in about 3500 BCE, an urbanized people who spoke a language similar to Turkic languages called Sumerian. Other groups of people speaking a Semitic language (those sharing cognates with Hebrew, Aramaic, and Arabic) lived just north of the Sumerian areas in Akkad; sometimes they moved into Sumerian cities, making life polyglot. But for several thousand years the Sumerian speakers predominated. We know this because they were the first to devise a way of writing and to leave records that we can decipher. "Sumer" refers to the area of cities where people spoke Sumerian, from Baghdad to the Persian Gulf, from about 3500 BCE until the destruction of the city of Ur by the Elamites (Iranians) in 2004 BCE. The same geographical area, including more land to the northeast, is often called Mesopotamia (Greek for "between the rivers") (Fig. 5.2.).[4]

In about 3800 BCE the monsoon winds and rain shifted southward, and people in Sumeria had to arrange more irrigation for their crops to succeed. They did this by moving into cities and organizing the irrigation of the surrounding land.

The first of eight or so Sumerian cities to emerge was Uruk, called Erech in Biblical times and Warka today. It lies 150 miles (250 kilometers) south of Baghdad and 12 miles (20 kilometers) from the Euphrates. By 3400 BCE Uruk had become the largest permanent settlement up to that time. It contained two major temples, one to An, the god of the sky, and the other to Inanna, the goddess of love and procreation.

Sumerians believed in a universe controlled by invisible living beings, or gods. There were thought to be seven major gods and goddesses, who formed a council that decided what would happen to people. The four major gods were An (sky), Enlil (air), Enki (water and wisdom), and Utu (sun); the three major goddesses were Ki (Earth), Nannar (moon), and Inanna (love and procreation, also known as Lady of the Evening, Lady of the Morning, and Queen of Heaven). These beings were thought to have given people a set of immutable and universal laws and rules, called the *me*, which had to be carried out to satisfy the divine beings.

As ranked hierarchies appeared in urban living, people began to rank their deities, and elite gods appeared. As the sociologist Émile Durkheim first suggested, our thinking about how the universe operates often reflects the way our own society works.[5]

Each major city had one or more gods in residence in divine house-holds, or temples. Temple statues were carved to embody the invisible divine spirit, and the temple staff worked to provide everything that its god or goddess could desire, so that it would choose to reside there and assist people. The temple staff controlled great acreage for food and tribute to keep their divine household well stocked. The main temple of each city stood on a high terrace, which gradually rose into a massive stage tower, or *ziggurat*, Sumer's contribution to religious architecture.

Over time Sumer changed from temple-dominated, fragmented cities into a centralized state, in which one city and its ruler controlled the oth-ers, backed by a bureaucracy of scribes and priests. As warfare became common among the cities, temple households became subordinate to war-rior households. Sargon of Akkad is Sumer's best-remembered king (ruled ca. 2350 BCE and for about fifty years); his grandson, Naram Suen (ruled 2291–2255 BCE) was the first to proclaim himself divine. Sargon achieved a new stage in state formation—a state that controlled several others— which he did by conquering others, demolishing their walls, and appoint-ing his sons as governors.

The food-producing land in Sumer was used in three distinct ways: as gardens within the cities, as irrigated fields lying parallel to the rivers, and as arid grazing land. The chief irrigated crops were dates, barley, wheat, and lentils, bean, and peas. Flax was grown for making linen cloth. Large herds of goats and sheep were kept, as were smaller ones of cattle for milk and meat, and donkeys and oxen for draught animals. Fish were a major supplement; poor people lived on a diet of barley, fish, and dates. Some hunting of rabbits and birds went on; dogs were ubiquitous.

Irrigation dominated the work of many people. In spring the river water had to be kept from flooding, then gradually released as the season wore on. This required constant repairs and adjustments to the dikes and canals. Since even fresh water has some salt content, salt crystals built up in the soil from the evaporating water, undermining the crops after several centuries.

At the fall equinox, after the harvest of crops but before the planting cycle began anew, Sumerians celebrated their new year with a temple ceremony at which the ruling king had sexual intercourse with the high priestess representing Inanna before the assembled public, to insure fertility for the coming year. The Sumerian word for water was *a*, which also signified sperm, or generative power. Sumerians clearly understood that life would not bloom without the masculine element.[6]

The earliest written language that has been found so far comes from the temple to Inanna at Uruk—on clay tablets incised in wedge-shaped patterns. This writing is believed to have developed from merchants pressing little tokens representing traded items into wet clay to record transactions. Later, officials drew pictures of the items in the clay with a stylus pointed on one end and ball-shaped on the other. Later, they let the pictures represent one-syllable names of the items. Someone decided to put a wedge shape on one end of the stylus, and the pictures began to be rendered by wedges. Thus emerged cuneiform writing (from Latin *cunens* for wedge), with about 3,000 characters representing syllables. This was used in Sumer and neighboring places for over 3,300 years.[7]

The Sumerian tablets that have been found so far, about 5,000 to 6,000, are scattered throughout museums the world over. An American team excavated at Nippur, Sumer's spiritual center, between 1889 and 1900; their findings are divided between the Istanbul Museum of the Ancient Orient and the University Museum of the University of Pennsylvania. The linguistic analysis of this material has been a triumph of scholarly cooperation, beginning in the late nineteenth century. New texts can now be translated with reasonable confidence, although many pieces of texts are still missing and could yet be found in Iraq when conditions permit. Most of the literary texts have been published, many in the last thirty years—among them twenty myths, nine epic tales including the *Epic of Gilgamesh* discussed in the last chapter, and several hundred hymns, laments, and dirges, including the hymn to Inanna at the end of this section. About 300 people today read cuneiform. Johns Hopkins University has established a project, called Digital Hammurabi, whose goal is creating an electronic archive of all known tablets in 3-D images so that scholars around the world can work on translating them.[8]

Since Sumerians had no timber, stone, or metal, they traded extensively by donkey caravan and possibly by boat with Turkey, Iran, Syria, the Indus Valley, and probably with Egypt. These connections formed a network of human communication in the core area of central Europe, Asia, and northern Africa.

At the beginning of city life in Sumer, copper was in use, when they could trade for it. Beginning in about 2500 BCE, people somewhere in western Asia learned to make bronze, a stronger metal made by combining one part tin to nine parts copper. Since tin was available only from Egypt's eastern desert, from Cornwall, England, and from Afghanistan, not much bronze was used for a long time. But bronze was being used by 2000 BCE in eastern Asia and by 1500 BCE in northeastern Africa.

The first human experiment with city life produced a frenzy of creative adaptation, in which many aspects of urban life emerged that remain today. The Sumerians developed a code of law and order—the *me* mentioned earlier—which they wrote into prescriptions of how conflicts should be resolved. Whichever ruler could organize the most effective army, bureaucracy, and supporting structures conquered the rulers of the other city-states. Wealthy families developed individual ownership of property and traded abroad for luxury goods. About 90 percent of the people remained farmers, sending tribute, coerced when necessary, to the rulers in exchange for protection. Ranked hierarchies, or distinct social classes, emerged, including slavery of failed farmers, nomads, and war captives. Sumerians invented pictographic writing, literature, cylinder seals, canals and dikes, weighted levers for lifting water, accounting procedures, entrenched bureaucracy, and the use of silver as money. Other places may also have invented some of these items, but the Sumerians put them all together.

Eventually, in 2004 BCE, 4,000 years ago, the Sumerian ruling city of Ur was destroyed by Elamites from Iran, and its king was taken into exile, never to return. Sumer's language died out, although cuneiform writing continued to be used as the language of international diplomacy until the first century of the Common Era.

Speculation about the reasons for Sumer's abrupt collapse has focused recently on the hazards of irrigation; increasing salinization resulted in decreasing yields and a series of poor harvests. Recent climate studies show that a major volcanic eruption to the north occurred in 2200 BCE, one that spewed ash sufficient to veil the sun. A 278-year drought cycle also began at the same time. Life in the first cities proved vulnerable to environmental changes.[9]

Sumer is with us still, in small specific ways as well as the larger ones just described. People of Sumer used a counting system based on twelve rather than ten. We retain this in counting our sixty-second minute, sixty-minute hour, twenty-four-hour day, twelve-month year, and 360-degree circle. We also retain the Sumerian belief in thirteen as an unlucky number and their belief in invisible ruling spirits, though we have reduced them to one.[10]

Listen to the joy of Sumerians as they sing their hymn to the evening star, Venus, representing Inanna, the goddess of love:

> At the end of the day, the Radiant Star, the Great Light that fills the sky,
> The Lady of the Evening appears in the heavens.
> The people in all the lands lift their eyes to her.
> The men purify themselves; the women cleanse themselves.
> The ox in his yoke lows to her.
> The sheep stir up the dust in their fold.
> All the living creatures of the steppe,
> The four-footed creatures of the high steppe,
> The lush gardens and orchards, the green reeds and trees,
> The fish of the deep and the birds in the heavens—
> My Lady makes them all hurry to their sleeping places.
> The living creatures and the numerous people of Sumer kneel before her.
> Those chosen by the old women prepare great platters of food and drink
>   for her.
> The Lady refreshes herself in the land.
> There is great joy in Sumer.
> The young man makes love with his beloved.
> My Lady looks in sweet wonder from heaven.
> The people of Sumer parade before the holy Inanna.
> Inanna, the Lady of the Evening, is radiant.
> I sing your praises, holy Inanna.
> The Lady of the Evening is radiant on the horizon.[11]

After the fall of Ur, plundering and raiding became consistent features of Mesopotamian life. Desert people moved in, and power shifted to Ham-

murabi (ca. 1792–1750 BCE) of Babylon. Babylonians fought back and forth
with Assyrians to the north, alternating conquests. The rule of Nebuchadnezzar II (604–562 BCE) is remembered, for he captured Jerusalem, destroyed the temple, and carried out a massive deportation of Judeans to
Babylon, known as the Babylonian captivity.

Sumerians traded, by coastal shipping and by overland caravans, with
two other early urban areas, the Nile Valley in Egypt and the Indus Valley in
Pakistan, probably from early in their development. The fourth urban area,
centered in China, remained separate from the core of Eurasia until later.

Other Urban Cultures—India, Egypt, and China

People began living in the Indus Valley in about 7000 BCE and on the river itself by 3000 BCE. They added humped zebu cattle and domesticated cotton
to the human resource base. Two Indus Valley cities have been excavated—
Mohenjo Daro and Harappa. Apparently these cities, arising in about 2600
BCE, managed water skillfully, separating drinking water from wastewater
in the first known sewage system. Since scholars have not been able to decipher the Indus script, nothing is known about their religion or government. Images carved on a few cylindrical seals indicate that some Hindu
gods may have originated as Indus deities. In about 2000 BCE some kind of
collapse began; gradual deforestation, salinization from overirrigation, invasions from the north, or shifts in the river systems seem the most likely
explanations. Something sudden, such as an earthquake or severe flooding,
may have precipitated the failure. By 1500 BCE urban life on the Indus River
had disappeared, demonstrating that turning to agriculture for food did
not guarantee a consistent production.

Much more is known about Egyptian society, because Egyptian hieroglyphics have been deciphered. Unlike the writing of the Sumerians, which
shows a gradual development, Egyptian hieroglyphics appeared as a full-
blown system, in about 3300 to 3200 BCE, suggesting a possible imitation of
the Sumerian system. Egyptian hieroglyphics were decoded in 1824 by
Jean-François Champollion, who used the Rosetta Stone found in Egypt by
Napoleon's troops. The inscription on the stone, from the second century
BCE, featured the same text written in three scripts: hieroglyphics, demotic
(a simplified hieroglyphics), and Greek. Most Egyptian texts are found on
papyrus, which preserves well under dry conditions. Papyrus roll books in
Egypt date from about 2500 BCE. Papyrus documents reveal that urban set-
tlements along the Nile were united in about 3100 BCE as a single complex

society governed from Memphis on the delta. Rulers, called pharaohs, declared themselves divine from an early time, another idea that may have come from Sumer.

The Nile River—the longest river in the world at 4,160 miles, with floods more reliable than any other river—presented advantages available nowhere else in the world. It provided reliable transportation by boat in both directions (the current flows north, the wind blows south), making it possible for the pharaoh to control shipping and distribution in his kingdom. The river provided annual flooding, which people trapped behind dikes for its deposit of soil, then released to water plants, avoiding the evaporation that caused salinization in Sumer. Thus the Nile provided the basis for an unusual stability, while the surrounding deserts provided a natural defense.[12]

The basic food crops in the Nile Valley were wheat, barley, dates, figs, olives, and grapes. Egyptians domesticated small fowl—ducks, geese, quail, pigeons, and pelicans—and caught many fish. They learned to make olives edible by soaking them in salt and water. They made bread and beer and used salt to preserve fish, which by 2800 BCE they were trading to the Phoenicians in exchange for cedar, glass, and purple dye made from the murex shellfish. Egyptians also learned to preserve human bodies by covering them in salt for seventy days.[13]

The religious ideas of the Egyptians cannot be systematized for lack of sufficient evidence. But a few basic assertions are clear. Egyptians were proud and judged themselves superior to others, a common human trait. Their population was mixed, composed of many different groups from Semitic peoples to dark black Nubians, and their gods were a blend of many local ones. Their creator god, Atum, was bisexual (the he-she god) and came to be associated with Ra, the sun god. Egyptians considered the heart to be the seat of intelligent thought and formulated a belief in an afterlife that one earned by thoughtful and moral behavior in this world. Osiris, the god of the dead who himself had risen from the dead, presided over the judgment, at which the newly dead person's heart was placed on a scale to measure whether the person would be snatched by the demons of death or proceed to an afterlife better than this one. Osiris's wife, Isis, was widely worshipped even outside of Egypt, especially in the early years of the Roman Empire.

The Egyptians managed to sustain their irrigation system for 5,000 years, longer than either the Sumerians or the Harappan society in the Indus Valley. Today, however, Egypt is beset with soil and water problems, since the technology of the twentieth century, applied to fix the problems, has worsened them. (For example, the Aswan Dam does not permit the annual flooding that deposited fertile silt; the dammed water is leaking underground into ancient tombs.[14]

The Egyptians, who did not feature much warfare among themselves, eventually suffered an invasion from people known as the Hyksos, thought to be the Canaanites from Palestine. In 1678 BCE the Hyksos used horse-drawn chariots, newly perfected, to cross the Sinai Desert and drag the Egyptians into the maelstrom of warfare that occurred between 2350 and 331 BCE among the rulers of the Middle East. Rulers everywhere had become skilled at building empires and maintaining them with a bureaucracy; by about 1550 BCE Egyptians ruled the Nile as far south as upper Nubia and the coast of Palestine and Syria north to the Euphrates River. Eight hundred years later Assyrians ruled Mesopotamia, and 200 years later the Persian Empire included lower Egypt, all of Turkey, and Mesopotamia to the Black and Caspian seas, stretching east to the Indus Valley. These military triumphs were supported by improvements in fighting, specifically horse-drawn chariots and iron metallurgy for cheap armor, starting in Cyprus or eastern Turkey about 1200 BCE (Fig. 6.1).

Egyptian culture strongly influenced a "country cousin" culture, the Minoan, on the island of Crete, located off the coast of Greece. Minoan culture was the first complex society of Europe, developing from 3000 to 1450 BCE. It used an early form of writing and deployed ships to found colonies and create a trading empire. Since Crete lies in the Mediterranean between Greece and the coast of Africa, Egyptian influences on Minoan frescoes, and presumably on the rest of its culture, were strong. Through Minoans, Egyptians influenced the Greeks, perhaps even providing analogs to its gods and goddesses. Minoan culture came to an abrupt end for unknown reasons, probably including the horrendous volcanic eruption on the nearby island of Thera, now Santorini, in 1645 BCE, which cast sun-reducing ashes into the atmosphere for years to come.[15]

In China, in the far eastern part of Eurasia, a fourth agrarian civilization arose producing another distinctive style of human culture. Early Chinese cities developed on the agricultural surplus produced within a great river system, the Yellow River and its tributaries. In northern China, where rainfall is more meager than in the south, millet (native) and wheat (spread from the Middle East) were the staple crops. Later, in wetter southern China, rice became the staple crop.

In China, urban areas evolved from well-established villages on terraced land near the Yellow River, unlike in Mesopotamia and Egypt, where cities grew on agricultural frontier land. By 3000 BCE there were walled vil-

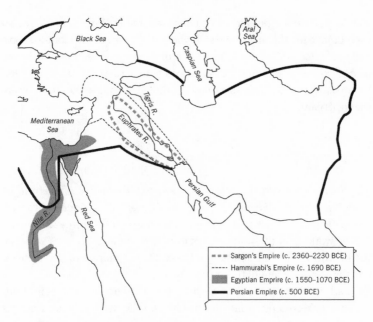

6.1   Some Ancient Empires of Southwest Asia and Egypt

lages in northern China with richly appointed tombs containing pottery with marks that appear to be ancestral to Chinese script. Elite families managed relations with the spirits, whom the Chinese believed could be reached through the spirits of their own ancestors interceding on behalf of their descendants. The ancestral spirits were contacted through offerings of alcoholic beverages in bronze vessels. The Chinese character for "ancestor" earlier meant "phallus" and even earlier meant "earth," suggesting the transition to a culture in which only the sons could perform the ritual sacrifices that set a father's soul free to join the ancestors.[16]

The same families who arranged contact with the spirits also organized defense, unlike the dichotomy between priests and soldiers in Mesopotamia. By 1523 BCE the Shang family had established military and political power by importing an expensive system of weapons from the Middle East—bows made of wood, bone, and sinew glued together, bronze armor, and horse-drawn chariots. The Shangs ruled for 500 years; their capital, Anyang, now in Honan province, has been excavated after inscribed bones kept turning up in farmers' fields. These bones were inscribed to cast oracles; the inscriptions are so close to China's historic script that scholars could read them at once.

During the years of Shang rule the Chinese elites used bronze with outstanding artistry, especially in ritual vessels and cooking pots. They also used bronze for the metal parts of wheeled vehicles but seldom for tools and implements. They made books of crossed bamboo sections and began using brushes for writing. They practiced human sacrifice and slavery and began using cowrie shells as currency, although no one knows from where the shells came.[17]

Urban Turning Points

The rise of the earliest cities in Eurasia and the Nile area of Africa brought many transformative changes to human life that we continue to live with today. As society grew more complex, certain structures seemed necessary for the working of the whole. Among these essential structures were the use of writing, the portability of religion, the elaboration of bureaucracy, and the establishment of the patriarchy.

Early writing proved immensely useful for religious activities, for trading, and for recording tribute. In the long run the elaborate arrangements of wedge shapes or pictographs proved unwieldy; people needed something easier.

In Egypt the pressure for a simplified system of hieroglyphics produced the demotic script, so named because it was the writing of the people. But the leap to an alphabet of letters, each of which represented a sound, was made by the Phoenicians, a Semitic people who were seagoing traders from the eastern end of the Mediterranean Sea (present-day Lebanon). Linking Egypt and Mesopotamia, they also founded Cadiz in southern Spain and from there sailed to the west coast of Africa in about 600 BCE, more than 2,000 years before the Portuguese were able to. Presumably their far-flung trading activities provided the impetus for them to create a simpler system of writing.[18]

In an alphabetic system, each symbol represents a single letter rather than a whole syllable. All the sounds of most languages can be represented with only twenty-five to thirty symbols. The Phoenician alphabet, developed in about 1400 to 1000 BCE, represented only the consonants, with twenty-two symbols or letters borrowed from Egyptian hieroglyphs (Fig. 6.2). This worked because Semitic languages had a limited number of vowels. Phoenicians consistently read their letters from left to right, unlike in Aramaic and Hebrew, two alphabets that developed in the eastern Mediterranean slightly later, which read from right to left.

An example of how our alphabet carries the continuity of history can

Phonecian Letters	Related Roman Letters	Phonecian Letters	Related Roman Letters	Phonecian Letters	Related Roman Letters	Phonecian Letters	Related Roman Letters
𐤊	A	I	Z	⅄	M	Φ	Q
𐤔	B	H	H	𐤉	N	𐤒	R
𐤂	G,C	⊗	–	‡	X	W	S
◁	D	ߙ	J	O	O	+	T
𐤀	E	⅄	K	𐤐	P		
Y	F,V,U W,Y	𐤋	L	𐤓	–		

6.2   The Phoenician Alphabet

be seen in the letter M. The ancient Egyptians drew wavy lines to signify water. This symbol is retained in the Hebrew and Phoenician letter "Mem," representing *mayim* or water, which became the Latin letter M.[19]

Creating an alphabet by making a phonetic analysis of the spoken language proved immensely difficult. Evidence suggests that it occurred only once in Afro-Eurasia and never in the Americas. Most alphabets borrowed from earlier ones or people got the idea of an alphabet from somewhere else and devised their own script.

By about 800 BCE the use of the Phoenician alphabet had spread to Greece, where people spoke a language with more vowels. Needing more letters for their vowel sounds, Greeks took letters for four extra consonants—A (alpha), E (epsilon), O (omicron), and Y (upsilon). I (iota) was a Greek innovation. Romans adopted the Greek alphabet, which is still used by Roman and Germanic languages.[20]

Arabic script also derives from the Phoenician alphabet, but it diverged from the Phoenician by about the start of the Common Era and turned into Arabic by the mid–sixth century CE. The Quran was written in Arabic script in 650 CE, and the script was widely spread by the rapid expansion of Islam throughout the world.

Chinese writing never abandoned its pictograms and syllabic characters. The Chinese system was invented in about 2000 to 1500 BCE, was simplified from 200 BCE to 200 CE, and remains essentially unchanged today. It uses about 214 keys, which have to be combined into characters that represent whole words.

Alphabetic writing proved transformative because it simplified reading and writing and opened it to a larger segment of the population. In the process it rendered sacred scriptures accessible to laypeople. The content of religious thought, long attached to local gods, became moveable as people could carry scriptures with them when they migrated or were captured. Local gods could morph into a universal god, no longer tied to a specific place.

The Israelites, or Jews, of Judea provided the dominant example of this process. Originally from Ur in Mesopotamia, Abraham led his family southwest into current Israel in about the twentieth century BCE. In 586 BCE the Babylonian king, Nebuchadnezzar, captured Jerusalem and destroyed the temple and its priests. The captured Israelites were carried off to Babylon, where they used their collection of sacred texts to construct a new kind of religion, one centered on weekly gatherings to hear the text explicated by teachers known as rabbis. By meditating on the texts, they constructed a code of conduct for the exiles and affirmed that God was universal—present wherever his people were and not residing in a particular place. Judaism has continued for 2,500 years to guide its believers wherever they live, under whatever hardship.

Alphabetic writing also made possible the elaboration of the bureaucratic structures necessary to maintain the empires that constant warfare in southwest Asia produced. Bureaucracy arose before the development of alphabets; it was firmly in place by the time of Hammurabi, who ruled Babylon in about 1792 to 1750 BCE. Bureaucracy meant that individuals, appointed by the ruler, had the authority to collect tribute and enforce laws, which people had to accept (most of the time) in return for military protection. Alphabetic writing greatly extended the effectiveness of the ruler's appointees. It also facilitated trade, as private people were able to record their own business contracts and transactions.

The rise of cities coincided with the establishment of the patriarchy, or the political and social subordination of women—another aspect of the development of hierarchies in human society. This cannot be explained by a single causal factor—as usual in complicated historical situations—but by a complex network of contributing factors all interacting to produce an outcome that characterized early urban societies.

During the transition to farming, women's role became more centered in the home. As men decreased their hunting, they used heavier plows than women could manage to increase their acreage. (But the patriarchy developed in the Americas in the absence of plows.) Increased food allowed more frequent babies, which kept women busier at home. Individual ownership of property led men to control women more, to ensure that their property went only to their heirs. Raids from outside groups made defense imperative; men had to organize the defense of their property and their families. Perhaps the simplest explanation is that men could be spared more easily than women from the most fundamental unit of society, the household, to specialize in other tasks.[21]

As cities developed and urban populations were no longer farming the land, the great mother goddess of early farming times began to lose meaning. The story of her overthrow can be seen in many mythologies. For example, the Babylonians told of their god-king Marduk, who waged war on Tiamat, the mother of all things. He hacked her body to pieces and fashioned the world anew from the pieces. The Israelites rejected the goddess image entirely. Their enemies, the Canaanites, a farming people, worshipped a fertility goddess known as Astarte, whom the Old Testament refers to as "the Abomination."[22]

In Greek and Roman cultures, the message of the power of men was made clear. Zeus gave birth to Athene from his head, a classic reversal of the great mother myth. Much of early Greek literature tells the same story of reducing the power of women. In *Eumenides*, Aeschylus has the sun god Apollo announce: "The mother is not the parent of that which is called her child, but only the nurse of the new planted seed that grows. The parent is he who mounts."[23] Boundary markers, called herms after the god Hermes, consisted of a man's head carved in the round at the top of a wooden or marble stake, on the front of which was added a set of male genitals, usually erect. These have been found in Greek culture from at least the sixth century BCE. In Roman culture the male sex organ was believed to have the power to avert and overcome evil influences. Phalluses were worn as protective amulets. Patriarchy was fully in place.[24]

In the approximate period covered by this chapter, 3000 to 1000 BCE, the world's population increased from about 50 million to about 120 million. Each century's rate of growth stood at about 4.3 to 4.5 percent, a gentle but not explosive acceleration. The long-term trend masks the cycles of expansion and decline that historians believe underlay the general upward trend.[25]

Some trading took place among the early cities, as documented by artifacts found far from their sites of origin. By 1100 to 800 BCE Phoenicians dominated trade in the Mediterranean, venturing down the west coast of Africa and to England for tin. By about 1500 BCE unknown metallurgists,

perhaps in the Caucasus, learned to smelt iron ore by raising the temperature of the furnace 400 degrees Celsius higher than required for smelting
copper. By 900 BCE iron tools were frequent in the eastern Mediterranean,
and during the first millennium BCE they spread through many regions of
Afro-Eurasia.[26]

From 3500 to about 800 BCE peoples in the core Afro-Eurasia network
began to develop the social systems and structures that would enable them
to sustain stable cities and large-scale empires. For all the local differences,
the solutions to dense living found in four areas of Afro-Eurasia turned out
to be remarkably similar to each other—and to those of agrarian civilizations that developed independently in the Americas (see chapter 10). Some
observers notice that these solutions are also remarkably similar to those
of termites and other social insects. Dense living may have its own characteristics whether among humans or insects.

In the next chapter we will see how, from about 800 BCE to 200 CE, the
systems and structures of early agrarian civilizations gave rise to what we
know as world religions and cultures. During those years the creation of
urban-based empire-civilizations reached its apex across the core of Afro-
Eurasia.

Unanswered Questions

1. Does the culture of civilization develop in one place and then spread to
other places?

Called a diffusionist view of culture, this idea was widely held fifty
years ago. But the view that culture arises in many places, only one in each
area, resulting in many pathways to civilization is the one most widely held
today.[27]

2. How much did Egypt trade with other Mediterranean settlements
and how much did Egyptian practices influence other cultures?

These questions have been much discussed in recent years as African
peoples have been eager to reclaim their contributions to world history,
particularly around the 1987 publication of Martin Bernal's *Black Athena:
The Afroasiatic Roots of Classical Civilization.*[28]

Bernal says that his central thesis in this book is that Egyptian and
Phoenician influences on the formation of Greek society were strong and
that these influences have been downplayed by European scholars for
racist and anti-Semitic reasons. Many historians concede that Bernal exposed many examples of how the influence of Egypt and Phoenicia has

been downplayed in the last 200 years, but most do not conclude that Greek culture was a construct of black African and Phoenician origin.

Aside from the question of how much influence Egypt had on Greek and Roman culture, there is the question of how black the Egyptians were. Were most of them black or Semitic or mixed? Since Egyptians used colors symbolically in their paintings, black-painted people may not actually have black skin color. There must have been a variety of skin colors from many different areas, but no one really knows.

7

The Afro-Eurasian Network

(800 BCE–200 CE)

The first rise of cities and civilizations on the planet occurred, as we have seen, in four large river valleys of Afro-Eurasia. These cities formed a network, a small core of urban life, in which people were trading and communicating with each other. During the period 800 BCE to 200 CE these societies elaborated the bureaucratic and religious systems that persevered as distinct world cultures into the global age.

Before our story moves on to what people in the core cities were creating, it is well to remember that outside and around the urban areas most people continued to live in various kinds of pre-urban life—in farming, herding, or hunting-and-gathering cultures. The Sahara Desert cut off the herding peoples of sub-Saharan Africa from the mainstream trade. People in the Americas continued their hunter-gatherer culture. North of the Afro-Eurasian core of rising urban civilizations lived the Celts of Europe and the horse-riding nomads of inner Asia. These nomads periodically swooped down into farming and urban areas; since they played a decisive role in the development of the subcontinent of India after early urban life there collapsed about 1500 BCE, they will appear briefly in this chapter but will be featured in the next two.

The Celts deserve special mention because their culture has often been downplayed in narratives of world history, since they were conquered by the Romans. At their height in about 300 BCE, however, Celtic people were living across Europe, from Ireland to the Black Sea, from Belgium to Spain and Italy, after originating in about 1000 BCE in the area of eastern France and western Germany where the Rhine and the Danube rivers begin.

Romans called the Celts "Gauls," from a Greek word, *hal,* meaning salt, for salt and iron were the basis of their economy. By 900 BCE iron tools and weapons were in frequent use in the eastern Mediterranean and Europe, where the Celts traded them up and down the rivers to which they gave names: Rhine, Main, Neckar, Ruhr, and Isar.[1]

The Celts lived a sophisticated communal subsistence farming culture, with elected officials, women's equality, excellent roads made of oak beams, stone buildings, exquisite metal jewelry and tools, and a lunar calendar more nearly accurate than the Julian (Roman) calendar. They worshipped many gods and goddesses led by their priests, the Druids. They used Greek letters for commerce but refused, in order to retain their capacity for memorization, to put their history, genealogy, and religion into print. Superb warriors, they preferred hand-to-hand combat and sometimes fought naked. They were able to pillage Rome for seven months in about 390 BCE, a little more than a hundred years after senators established the Roman republic. But the Roman Empire fought back, as will be seen, eventually conquering all of Celtic territory except Ireland, Wales, Scotland, and Brittany (the north coast of France), where the Celts have been able to maintain their 3,000-year-old culture into modern times.[2]

India

Turning to the urbanized areas of Afro-Eurasia, we start with India bounded on the north by the Himalayas, which were created when India was a separate floating continent that crashed into the Asian mainland. Only one passable route through these majestic mountains existed—the Khyber Pass. In about 1500 BCE, horse-riding nomads from the north, speaking an Indo-European language, migrated into northern India through this pass. Did they attack and conquer or quietly assimilate? The answer is not known.

They did arrive, however—these people known as the Aryas, lighter-skinned than the indigenous Dravidians. Somehow the two groups merged into a system of *varnas* (literally colors) or castes, in which the lighter-skinned Aryas occupied the higher castes and the Dravidians the lower ones. Priests and scholars held the highest rank (Brahmins), then the warriors and rulers (Kshatriyas), then all the other Aryas, and finally the non-Aryas. The caste system became distinctive of Indian culture, perpetuating itself by marriage strictly restricted to fellow caste members. Today the caste system has been officially abolished, but there are still 25,000 subcaste groups organized into 3,000 castes, classified into the four ancient varnas.

The caste system became associated with a belief in reincarnation. The priests taught that every living creature has an *atman*, or self, that at death separates from the body and returns in another body, depending on its karma, or deeds done during its life. If the self and its body accepts the caste position allotted to it and carries out its duties faithfully, then the self is rewarded by returning in the next life to a higher caste. If not, the self is punished with a lower caste. This belief system helped people accept the socioeconomic immobility of the caste system.

The people of India enjoyed little political unity, since local rulers were unable to solicit enough support from across their castes to conquer neighbors and since nomadic groups kept arriving from the north to raid and to assimilate. Because horses did not fare well in India's climate, Indians could not resist the horse-riding warriors from the north.

Indian religion proved tolerant and gradually added local gods until it became a conglomeration of multiple gods and goddesses—330 million, according to one tradition. Yet within this multiplicity was unity; all the gods and goddesses were considered manifestations of a single divine force pervading the universe. The name by which this religion is known today emerged in about the eleventh century CE, when Islamic invaders called it Hinduism, meaning "what Indians do."[3]

Eventually various beliefs and practices developed that challenged both the authority of the Brahmin priests and the idea of endless reincarnation. Yoga was one of these practices—the idea that individuals could gain liberating insight through mental and physical discipline. But the most influential challenge came from an individual named Siddhartha Gautama (563–483 BCE), who became known as the Buddha, or "Enlightened One."[4]

Gautama was born the son of a king of a small kingdom in what is now Nepal, into the Kshatriya caste. After a regal upbringing and a luxurious life for twenty-nine years, he renounced his privilege and became a wandering ascetic. After six years he realized that asceticism was no more likely to produce insight than was luxury, and he chose a middle path. Seated under a banyan fig tree south of the present city of Patna in northeast India, he had a sudden and profound insight that became the basis of his teachings. He stressed living modestly, to minimize desire and suffering, and searching through self-discipline and meditation. He did not believe in gods or a single god nor in the persistence of a self or soul after death. His goal was to achieve nirvana, literally "snuffing out the flame," a release from the cycles of reincarnation. He lived out his sense of leadership by traveling throughout India to teach his insights.

The Buddha's teachings attracted many followers, who took vows of celibacy, nonviolence, and poverty. As his teachings spread, a schism developed between those who retained the original teachings (Theravada Buddhism) and those who added new teachings, such as worshipping the Buddha as a god and revering *bodhisattvas*, or people who almost attain nirvana but choose reincarnation in order to stay on Earth and help others.

The Indian subcontinent achieved political unity after the death of the Macedonian-Greek ruler Alexander the Great, who reached the Punjab (northern Pakistan) in 326 BCE. After Alexander's death, the Indian ruler Chandragupta Maurya, could extend his control. From 269 to 232 BCE the great king Ashoka further extended the kingdom by conquest. Full of remorse, Ashoka converted to Buddhism and practiced nonviolence, morality, tolerance, and moderation. He outlawed animal sacrifice, stopped the slaughter of animals in his kitchens, and gave up royal hunting trips. The Buddhist wheel of the law, adopted by Ashoka, still waves on India's flag today; under Ashoka Buddhism became a world religion.

About fifty years after Ashoka's death the government of northern India collapsed under attacks from the north, not to be reunited for another 500 years. Yet this period of feuding kingdoms from the third century BCE to the third century CE is considered a classical period of the flowering of Indian art and literature.

China

In China, there was no collapse of early urban life, and a distinctive culture based on appealing to ancestors continued. After the Shang dynasty, China endured a period of at least twenty-five warring feudal states vying for power from about 1030 to 221 BCE. Dikes and canals were built to bring the whole floodplain of the Yellow River under cultivation. Many new inventions became integrated into the culture, among them animal-drawn ploughs, the trace harness, crossbows, and a money economy. Bronze appeared in China in about 1500 BCE, while iron production started at about 500 BCE. The Chinese wanted jade from central Asia, while Mediterranean people wanted lapis lazuli from Afghanistan and Iran; hence, weak trading routes were in place between the Mediterranean and China during this time.

The trace harness, in use in China from the fourth century BCE, used a breast strap low across the horse's collarbone rather than across its throat, the latter of which reduced the horse's efficiency by partially choking it. Europeans probably did not invent the trace harness; central Asian people brought it to Hungary in the mid–sixth century CE; until then, horses in China could pull heavier loads than those in Europe.

To protect themselves from nomad attacks, the Chinese developed massive arms production, especially of crossbows, which could pierce two suits of metal armor. The trigger mechanism on the crossbow involved three moving pieces on two shafts, each cast in bronze and ground to precision. Greeks used crossbows in the fourth century BCE; whether they were smuggled out of China or copied, no one knows. Crossbows disappeared in Europe from 400 to 900 CE, then reappeared and were used by Cortez as one of his main weapons in subjugating the people of Mesoamerica.[5]

Despite the political instability in China, it enjoyed a period of intellectual development. Hundreds of schools of philosophers traveled about giving advice and setting up academies. The feudal system was replaced with a bureaucratic one, complete with police and passports. Coins stamped with their value appeared in the middle of the first millennium BCE.

Steppe nomads kept attacking China in the north and northwest; from them the Chinese learned chariot warfare and the use of saddles and stirrups. By about 350 BCE the Chinese had learned cavalry warfare, although horses were expensive to keep in a land without grass. (One horse ate as much grain as twelve people.)[6]

In 221 BCE the family of Chin managed to unify the whole Chinese empire. The first emperor, Chin Shih Huang Ti, proved indefatigable, poring over a ton of reports on wood and bamboo tablets per month. He took over estates and had them managed by the government. He standardized weights, measures, and the gauges of carts and chariots, and extended the Great Wall. Fifteen years later, however, power passed to the Han dynasty, which ruled China from 206 BCE until 220 CE.

The Chinese empire rested in part on the surplus of grain produced by its system of canals, which also provided the transportation for taxes to be collected. The tax was a percentage of the annual harvest, collected in kind and delivered to court. Men also owed one month a year of labor and two years of military duty. The walls of Changan (now Xian), the capital from 202 BCE until 8 CE, enclosed about 246,000 people in 2 CE, when the first national census was taken. About 60 million people lived in the empire, an estimated 10 to 30 percent of them in cities, compared to about 10 percent living in cities in Europe.

The other foundation of the Han dynasty lay in the moral teachings of

Confucius (Latinized form of Kong Fuzi, or "Master Kong"), who lived in about 551 to 479 BCE. Confucius taught the proper way to live—namely, that social hierarchy is a natural phenomenon and that gentlemen should cultivate good relationships with the spirits by behaving properly in private and in public. Under the Han rulers, the study of Confucian texts became the mark of an educated man and the qualification for holding administrative office, won through competitive examinations.

During the period 480 to 221 BCE, many among the masses of Chinese people began to follow the teachings of Laozi (or Lao Tsu), about fifty years older than Confucius, although the facts are not firm. Laozi advised renouncing worldly ambition, focusing on self-enlightenment and finding one's own path (the Tao) to right conduct, which his followers preferred to social, bureaucratic, and governmental requirements. Laozi's teachings became known as Taoism or Daoism.

Trade along the routes from China to central Asia and to the lands of the Mediterranean jumped to a much higher level in about 101 BCE when the emperor Han Wu Ti (Wudi) sent an envoy to fetch a special kind of large horse raised in the Fergana Valley, now Uzbekistan. This envoy, Zhang Qian, made eighteen trips altogether, although the first took thirteen years round-trip because he was captured on the way. His route became known as the Silk Road, after China's main export. Made from caterpillars spinning cocoons in mulberry trees, silk was a secret the Chinese kept until the sixth century CE. Silk became a kind of currency and the most important form of property in central Asia. Greeks and Romans treasured it; Buddhists needed large quantities for banners. Seeds and crops were also exchanged along the Silk Road; wine grapes and alfalfa went to China, which sent apricots and peaches to the Mediterranean (Fig. 7.1).[7]

During the Han era a skeptical and rationalist way of thinking developed in China, just as it had in Greece a century or two earlier. An active intellectual life went on, including the invention and spread of paper. Imperial officials kept registers of land and households in order to keep track of money and services due. Chinese textile technology was not approached by Iran or Europe until centuries later. Coal was used as fuel for iron making.[8]

The loads transported on the Silk Road hid invisible travelers—the viruses of animal diseases. Some of these diseases we still know as childhood illnesses—smallpox, mumps, whooping cough, and measles. These diseases did not spread from animals to people until people were living in densities of approximately 300,000 or more, insuring a constant source of new hosts for the virus after the death of its first.[9]

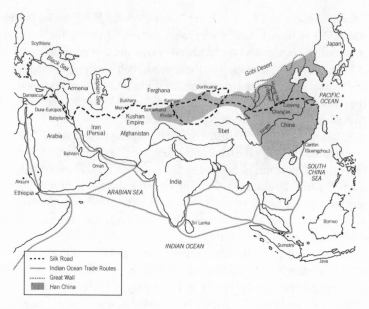

7.1   The Silk Road and Han China

The Silk Road spread these diseases to vulnerable city populations at both its ends. From about 165 to 180 CE serious epidemics occurred in both the Roman and the Chinese empires, claiming the lives of up to 25 percent of the population, one factor leading to the downfall of the Han dynasty in 220 CE.

The Han empire contended as well with intrigues in the ruling class, corruption and inefficiencies, uprisings of desperate peasants, the spread of banditry, and the ambitions of rural warlords. The underlying instability that the Chinese could never overcome, however, was the continuing raids and attacks of nomadic peoples from the steppes. After the Han regime fell, China experienced a period of political fragmentation that lasted until the late sixth century CE.

The Silk Road trade consolidated the network among Afro-Eurasian cities and agrarian civilizations. It linked China, India, Greece, and Rome in a heavy traffic of ideas and products that initiated a new era in world history, as we will see in the next chapter. This connection between China and the Mediterranean proved no less significant for the first millennium than the link to the Americas by Columbus did for the modern world. Before moving into this new era of the Silk Road link, however, we must catch up with life and culture around the Mediterranean Sea.

Greece

More is known about the Greeks than is known about early Mesopotamia, China, or India, because the evidence is more abundant, through excavation and preserved texts. The Greeks did not experience a rapid rise into dense urban regions with extreme social stratification. The rocky hills of the lower Greek peninsula, with their crops of barley, olives, and grapes together with sheep and goats, could not support large urban populations. Northern Greece had enough rainfall for raising horses and cattle. But the fallout from the volcanic eruption on Thera (Santorini) in about 1650 BCE influenced the climate for years to come. Not until about 800 BCE did the Greeks create an urban form called the *polis*, new in the world and one that incorporated more of the equality that the Greeks knew from tribal life than did cities in Mesopotamia, India, or China.

The polis, usually located on a hilltop, included the surrounding countryside. It was an association of citizens, led by a magistrate elected for a limited time, usually a year. Citizenship was open only to males, for it required fighting for the polis in person; women, children, slaves, and foreigners were excluded. Wise magistrates such as Solon in 594 BCE forgave debts and rearranged property and voting rights to prevent the gap between rich and poor from becoming too large.

In about 700 BCE alphabetic writing reached Greece from Phoenicia, enabling Greeks to write down the epics attributed to Homer. Greek colonists established hundreds of new cities around the coasts of Greece, Italy, Turkey, and the Black Sea, as the Greek population increased five- to sevenfold in the eighth century BCE. The colonists became the model of individualism, much admired by Greeks, who called themselves Hellenes and other people *barbaroi*, literally non–Greek speaking. People in the city of Lydia, in present western Turkey, invented coins with their value stamped on them as a way for the state to guarantee the weight and purity of gold, silver, and copper. The Greeks hastened to adopt coinage and, as a result, trade flourished.

In the *Iliad*, the Greek poet Homer honored individual valor in fighting. Yet by the mid–seventh century BCE the Greeks were using phalanx fighting on land, in which citizens fought side by side in lines, each using his shield to protect the man next to him. In this way the pursuit of fame and glory was transferred from the individual to the polis as a whole. In 480 BCE, in a surprise victory at sea, a coalition of about twenty Greek cities defeated the army of the Persian empire. A year later the Greeks defeated them again on land, inaugurating a golden period of 150 years of Greek cultural creativity led by its largest city, Athens.

During these years Athens held about 300,000 people, of whom 30,000 to 40,000 men were citizens. They developed a distinctive upper-class style consisting of political discussions in the *agora*, athletic games which men played in the nude, and drinking parties with philosophical debate, all based on developing individual power and using the mind to reason without any constraints set by priests or kings. The citizens practiced direct democracy, with legislative power vested in the whole body of citizens and executive power in a Council of 500, limited to two-year terms. Ten generals had to be elected annually but could serve indefinitely. Pericles served as a general from 461 to 429 BCE, the height of Greek democracy.[10]

Loyalty to the polis came first for the Greeks, and magistrates, rather than priests, arranged the religious rituals. A panoply of gods were worshipped by the upper class; led by Zeus, the sky god, and Ares, the war god, these deities behaved like humans, except that they enjoyed immortality. The agricultural majority of people practiced fertility cults based on female deities such as Astarte.

Teachers called Sophists provided training in logic and public speaking for upper-class men. In the absence of an authoritative priesthood, Greek thinkers applied their power of verbal reasoning to all areas of life. Drama, poetry, history, philosophy, and science flowered, reaching a climax in the questions raised by Plato (d. 347 BCE) and the answers given by Aristotle (d. 322 BCE).

Athenian women seemed to have lived under complete patriarchy, but their experiences differed from city to city, and even the patriarchal model may be more men's idealization than a description of historical conditions. Sparta's aristocratic women acquired some wealth and autonomy when their mercenary husbands were off soldiering, but some men in Athens, including Aristotle, thought the Spartan women were licentious, greedy, and the cause of Sparta's decline.[11]

During these golden years about one-third of the people of Attica (southern Greece) were slaves. Most of the slaves were foreigners; some were natives fallen into debt. Slaves did any work and submitted to any sexual acts ordained by their owners, but there seems to have been little extreme cruelty or abuse. The slave Epictetus became a philosopher still read today. Greek writers justified slavery by claiming that the barbaroi were not as rational as the Greeks and hence were better off under their care.

Why was Athens so successful for a period of time? Whatever else, the city was unusually rich, gaining great wealth from silver mines in its territory. It also functioned as a mini-empire, collecting tribute from other city states in the league it established to defeat the Persian navy. By its expansionist policies Athens provoked the Peloponnesian War with the other Greek states (431–404 BCE). Athens and its allies lost this war to Sparta and its allies, and the Greek states never figured out how to live in peace. Athens regained its freedom, destroyed Spartan expansion, and continued its Golden Age another half century.[12]

In 338 BCE, sixteen years before the death of Aristotle, Philip of Macedonia (the area just north of the Greek peninsula) conquered Athens and the other Greek cities. Two years later Philip was assassinated, perhaps at the instigation of his estranged wife. His son, Alexander, who had been tutored by Aristotle, took over at age twenty. Within fifteen years Alexander the Great conquered the Persian empire, creating the largest one known to that time. His conquests included Egypt and the northern coast of Africa. Greek ideas and style were incorporated throughout Alexander's empire as Greeks helped to administer it. Whatever Egypt had contributed earlier to Greek and Roman culture, the flow now reversed as Egypt became Hellenized, then Romanized.

During their heyday, Greek cities reduced the forest cover of Greek lands from about 50 percent in 600 BCE to about 10 percent in 200 BCE. They used timber for heating, cooking, firing pottery, smelting iron and bronze, and building ships. They used four-fifths of their land area as pasture for sheep and goats, resulting in severe overgrazing.[13] When conquered by the Romans from 215 to 146 BCE, the Greeks succumbed to the imperial, bureaucratic governance more common elsewhere in the urbanized world.

Rome

The Italian peninsula served as the focus of the other civilization on the northern side of the Mediterranean. Here geography proved friendlier than in Greece; Italy was more fertile and could support a larger population. During the seventh and sixth centuries BCE Etruscans from Etruria, centered on the western side of Italy between the Arno River (Pisa and Florence) and the Tiber River (Rome), dominated the peninsula until checked by the Romans.

In about 600 BCE seven hill towns in central Italy on the western coast merged to become Rome. In 507 BCE members of the senatorial class overthrew a ruling tyrant and established the Republic of Rome, which lasted until 31 BCE. Under the republic all male citizens could vote, but the vote of a wealthy citizen counted more than the vote of a poor one. Increasingly, hereditary senators governed as Rome institutionalized inequality. The oldest male member of the family, called the *paterfamilias*, held authority over all other family members.

During the time of the republic, Rome gained control of the Italian peninsula and acquired its first overseas colonies—Sicily, Sardinia, and Spain. Rome's most brilliant general, Julius Caesar, conquered the Celtic people of Gaul, now France, in 59 to 51 BCE. Every year a different senator served as governor in the provinces, but over time this system proved inadequate as Rome extended its control to the Rhine; to Vienna, Budapest, and Belgrade along the Danube to the Black Sea in Europe; parts of Turkey; the Near East; and northern Africa. In 47 CE the Romans conquered southern England but never Ireland, Scotland, and Wales. Here the Celts held the line.

People from the conquered areas were sold into Roman slavery in huge numbers, including 500,000 captured by Julius Caesar in his nine years of fighting in Gaul. There is no reliable evidence for estimating the overall proportion of slavery in Roman society, but Roman emperors had about 20,000 slaves and wealthy families had as many as 4,000.

In 31 BCE the republic ended as Octavian, the grand nephew of Julius Caesar, became Augustus, an all-powerful emperor-dictator who gave Rome an administrative bureaucracy capable of managing its empire with considerable honesty and consistency. The period 27 BCE to 180 CE is known as the Pax Romana, a period without major warfare in western Europe. The height of Roman domination occurred in the second century CE.

The city of Rome in the first three centuries of the Common Era contained about a million people. For food its people depended on grain shipped from Sicily and northern Africa. Roman culture blended many Mediterranean influences; Roman gods absorbed Greek ones, as Jupiter included Zeus and Mars included Ares. Roman roads connected over 5,000 miles from Scotland to Palestine; on them one could average about 92 miles a day by horse.

Salt proved a necessary ingredient in Roman success; more than sixty salt works were developed throughout the empire. The army needed salt for the soldiers and horses. Soldiers were sometimes paid in salt (the origin of the word *salary* and the expression "worth his salt"). Fish were the centerpiece of Roman cuisine and salted fish one of their chief commodities for trading.[14]

One of the satellite states of the Roman system was Judea, the Jewish state at the eastern end of the Mediterranean. The height of the Jewish state had come in the tenth century BCE, under kings David and Solomon. By 933 BCE the state had split into Israel and Judah; Israel fell to Assyria in 722, and in 586 Judah fell to the Babylonians, who destroyed Jerusalem and its temple and deported at least 10,000 Jews to Babylon. In the fifth century BCE Jerusalem was rebuilt but Judah remained a satellite state in some imperial system or other—Persian, Hellenistic, and, after 68 BCE, Roman.

In a fateful event in 6 CE, the homeland of the Jews (roughly equivalent to modern Israel) came under direct Roman rule. The Roman governors could tolerate the Jewish belief in one god only if images of Roman imperial power were displayed. This, together with high taxation, provoked Jewish resistance. Many Jews began to hope for the imminent arrival of the Messiah, the "Anointed One," who would drive the Romans out.

In this context Jesus, a young carpenter from Galilee in northern Israel, began to teach. He opposed the Jewish religious leaders, both Sadducees and Pharisees, whom he found excessively concerned with money and power. Urging a return to personal faith and spirituality, he was considered a political agitator and potential revolutionary by the other Jewish leaders, who had made their compromises with Roman authority. Jesus was handed over to the Roman prefect Pontius Pilate, who permitted Jesus to be condemned and crucified, a punishment usually reserved for common criminals.

Jesus's followers, believing that he rose from the dead after his crucifixion, spread his message of God's love. Paul, a Jew from the Greek city of Tarsus in Anatolia (Turkey), accepted the teachings of Jesus. From 45 to 58 CE Paul recruited followers in Greece, Syria-Palestine, and Anatolia, traveling on Roman roads, using the term Christian (from *christos*, Greek for "anointed one"), setting up Christian communities around the eastern end of the Mediterranean.

Back in Jerusalem things were not going well. Judea was ruled by a succession of Roman functionaries. Pontius Pilate was replaced in 36 CE. Tension between rich and poor, between city and country, increased. When work on the Temple Mount, begun under Herod the Great, was finally finished in the early 60s CE, 18,000 landless men were out of work. To keep social peace, the local government created the first known "make-work" project; men were paid to repave the city streets, even if they worked only an hour a day. A generation after the death of Jesus, in 66 CE, the Jews in

Judea finally revolted against the Roman rulers; their revolt was crushed
with the destruction of the temple in 70 CE, and with it the Christian community in Jerusalem. This cleared the path for Christians to diverge from
their Jewish roots, to become more Hellenized, and to become a sizeable
minority in the Roman Empire despite, or perhaps because of, governmental persecution. It left the Jews without a government of their own until the
establishment of modern-day Israel in 1948.[15]

Christianity created identity and community for the poor and op-
pressed of the Roman world, especially as the empire experienced its
process of disintegration. This began in about 165 to 180 CE, when the epidemics mentioned earlier wiped out about a quarter of the population of
the Roman Empire. Christians may have coped more successfully with
these epidemics than did other groups of Roman society. Christians recognized caring for the sick as a religious duty; non-Christians often fled and
did not provide basic food and water to the ill, some of whom would have
survived with basic care. Proportionately more Christians survived, and
those who did felt warm gratitude to their community. Christian teachings
made life meaningful even amid sudden death; heavenly existence was
promised for missing friends and relatives.[16]

In the third century CE the Roman Empire suffered severe monetary in-
flation. Food prices rose. Gold and silver sources were depleted. The government put less gold and silver into newly minted coins, people lost
confidence in coinage, and the economy reverted to barter and taxation in
kind.[17] Brigandage, always a sign of social distress, increased. The population shifted from cities back to countryside as people sought protection
on the estates of powerful landowners. For a brief period Constantine
(r. 306–337 CE) reunited the empire from Rome and converted to Christianity, but in 324 he transferred the capital to Byzantium, renamed Constantinople. In 410 Rome was sacked by invading Visigoths from the steppes of
central Asia. The empire continued from Constantinople, while Italy and
the rest of Europe fell into localized power structures.

The dissolution of the Roman Empire happened over many generations, not as suddenly as did those of Mesopotamia and Minoa. The complexity of factors contributing to such decay defies analysis. Historians
have pointed to moral decay, climate change, lack of military preparedness,
prevalence of slavery, and abuse and degradation of natural resources.
Reduction in agricultural production resulting from erosion and overgrazing certainly contributed. Urban people in Italy depended on far-flung
colonies for their food and way of life, and eventually the central authorities
could no longer maintain control of the colonies.

Greek and Roman upper-class patterns of citizenship and freedom lasted only a few centuries, but the ideas lived on orally and in texts preserved in libraries. When conditions were ripe again in Italy, these ideas would resurface to put their imprint on the whole European experience.

Population, Environment, and Religion

Since the world's first two censuses were taken during the Han and the Roman empires as part of their bureaucratic programs, no one knows how many people lived before those times or outside those empires. Historians estimate that as farming got started there were perhaps 6 million people living in the whole world. This population increased to 100 million by 1000 BCE and to about 250 million in 1 CE. Agriculture released the prior constraints on human numbers; with surplus food people could raise many children to adulthood. Innovations in technology led to population increases, which outstripped capacity and led to declines, in cycles. In 100 CE there were about 75 cities of 30,000 to 450,000, with a total large-city population of perhaps 5 million.[18]

The people added by agricultural success came at a cost, both natural and human. The natural cost consisted of the human degradation of the environment that sustained them—principally deforestation, soil erosion from deforestation, and salinization of the soil from irrigation. Deforestation, caused by the pasturing of sheep and goats and by the need for fuel for heating and cooking and for charcoal for pottery and metallurgy, is the background for the development of all human society.

These environmental impacts were strongly in evidence in Afro-Eurasia by 200 CE. The plains of Sumer had become completely barren. The complex society that emerged in the Indus Valley lasted only about 500 years due to deforestation and salinization. The loss of trees in China caused flooding by the Yellow River, which came to be known by the color of soil carried in it. The cedars of Lebanon, prized for their height and straightness, were a mainstay of Phoenician commerce; only a few small groves remain today. The Mediterranean shores lost their natural vegetation of oaks, beech, pines, and cedars; only olive trees would grow on badly eroded hillsides, their roots strong enough to penetrate the limestone rock. The Roman provinces of North Africa were reduced to vast deserts. These areas never recovered from the degradation of their environments from 800 BCE to 200 CE. Only Egyptians reached a sustainable balance, for 7,000 years, with a river that naturally renewed the soil downstream each year

(with soil eroded from up-river locations) until, in the twentieth century, irrigation and dams stopped the flooding and destroyed the natural cycle.[19]

There were costs to humans, too, from the increasing human density.
Life in cities and their outlying tribute areas proved difficult and precarious; many adjustments had to be made, and much misery had to be borne.
By 200 CE the vast majority of humans still lived in villages, but a growing
proportion of them had to pay tribute or taxes to urban-based rulers. Those
living within cities were divided into an extreme social stratification, between those owning land and the landless majority who provided their
labor for subsistence pay. Only a minority elite enjoyed the fruits of civilization. Warfare became a constant feature of life in order to extend control
and to protect stored food surplus. Finally, after travel on the Silk Road
linked dense populations across all of Afro-Eurasia, urban people had to
face the devastation of epidemics caused by animal viruses and bacteria
that could only spread to dense human populations.

During the thousand years of the period 800 BCE to 200 CE, Afro-Eurasia
witnessed a notable creative outpouring of religious thought. All of the
major world faiths of today appeared in this period, if one counts Christianity and Islam as developments of the prophetic stream of Hebrew life.
The great sages of Eurasia appeared during this era—Zoroaster[20] and Mani
(Manichaeism) in Persia, Confucius and Laozi in China, the Vedic seers and
Buddha in India, the Greek philosophers, the Jewish prophets, Jesus and
later Muhammad in the Mediterranean. Outside the urban areas—in the
Americas, in northern Europe, and in sub-Saharan Africa—people seem to
have continued their earlier religions.

The new religions that appeared in the urban core seemed different in
kind from earlier ones. During hunting-and-gathering days people re-
spected the invisible spirit world that paralleled all living things; later they
worshiped gods and goddesses who behaved like people on a grander scale
but without mortality. These were both life-affirming postures, with joyous
celebrations and fearful pleadings, but focused on valuing the given natural
world in which people were embedded. Pre-urban and non-urban people
did not experience much change during their lifetimes, and presumably
they valued the abiding quality of life—the sense of primordial time and life
as it always is.

The new religions in the period 800 BCE to 200 CE turned away from af-
firming this world, which no longer seemed wholly satisfactory, to envisag-
ing a better, transcendent world. The new prophets and sages stressed how
to gain salvation, release, or nirvana, how to achieve a better life after this
one, or how to return to a better incarnation. In part, they were working out

ethical systems that would motivate people to the behavior needed for living in the density of growing cities.

At the same time, the prophets and sages were finding antidotes for the personal distress that so many people experienced in city life—a psychic compensation for the misery and uncertainties of urban life or of adjacent village life, where people paid tribute to urban elites. Belonging to a religious community in a city gave people the opportunity to reestablish the kind of small, caring group that had characterized life in small bands of hunter-gatherer societies and in village life before the beginning of urbanization.

Living in cities had many advantages, especially for the landowning elites. But always it involved betraying the local customs of pre-city life. It presented a choice to people; whether to imitate civilized city ways with hopes for a better life, thus rejecting village custom and primordial life, or to repudiate civilized ways and strengthen traditional customs. This same choice confronts many people today, as the forces toward urbanization intensify to include every area of Earth.

Unanswered Questions

1. Historians and philosophers generally agree on the appearance of world religions in a certain period of time, from 800 BCE to 200 BCE, named the Axial Age by the German existentialist philosopher Karl Jaspers in a world history first published in 1949, called *The Origin and Goal of History*. What are the reasons for the roughly simultaneous appearance of the world religions?

There is no agreement yet on the answers to this question. Possibly these religions, with their emphasis on the next world, reflected the difficulty of life in the density of cities. Individual thinkers, for the first time, seem to have been able to separate themselves from the community consciousness and imagine their own answers; perhaps this resulted in part from increased contacts among people with different ideas. The presence of alphabetic writing, a vehicle for organizing and spreading the results of individual reflection and thought, must have played a significant role; Jews, Christians, and Muslims believe specifically in a book. Increasing travel and trade in networks of routes meant that sharing and collective learning increased. Questions about the nature, the causes, and the consequences of the so-called Axial Age are ripe for research and clarification. Whatever the reasons for the rise of these religions, many people are still living with thought systems created some 2,000 years ago, rather than with newer ones that might relate to current knowledge and conditions.[21]

2. Did the Roman Empire "fall"?

The conventional view in Europe and the United States has held that the disintegration of Roman military and political power marked the end of a civilization, leaving Europe in the grip of a dark age of material and intellectual poverty. In the 1970s Western historians began to avoid the words "decline," "fall," and "crisis," and instead used "transformation," "change," "transition." In the 1980s a German scholar, A. Demandt, in *Der Fall Roms*, made an alphabetical list of 210 reasons given over the centuries to explain the "fall" of the Roman Empire, while an American, Alden Rollins, reviewed briefly the many books over the centuries about the fall of Rome; both showed how writers could attribute the "fall" to almost any cause. Change does not necessarily mean decay, and, in any case, "fall" implies some speed, while the complex process of transformation proceeds gradually. With this discussion in mind, I have referred to the "dissolution" rather than to the "fall" of the Roman Empire.[22]

8

Expanding the Afro-Eurasian Network

(200–1000 CE)

By the beginning of the Common Era, humanity had achieved a poten-
tially stable pattern of life under conditions of urbanization. Powerful leaders had developed the imperial, bureaucratic command system as
the institution that tried to guarantee stability to its populations. The
Roman Empire, the Han dynasty in China, and the imperial structures in
Mesopotamia, Iran, and India all represent a climax of adaptations arising
from the shift to agriculture that produced urban living.

Severe human inequalities arose from this shift, however, and raiding, plundering, and warfare became a feature of life—the apparent cost of
the notable increase in numbers of people. Stability did not last. New developments in transportation and commerce arose to upset the imperial
command system and to inaugurate the next great leap in humanity's
development—the intensification of trade into powerful commercial
networks.[1]

The Central Core (200–600)

Europeans and Americans are accustomed to calling the period after the
demise of the Roman Empire "the Dark Ages." In this perspective the light
went out when Latin cultures wedded to Christianity dimmed. But, on the
other hand, the light went on for Angles, Saxons, Goths, Vandals, Franks,
and other groups in Europe, and it went on brightly for the empires located
in the center of Eurasia. Looking at the whole extent of the Afro-Eurasian
network of communication, we can say that with the demise of Rome and

the Han dynasty, at either side of the continent, the locus of imperial power shifted to Iran and India. Europe suffered more than China did and took longer to recover.

The withdrawal of power from Europe, when the Roman emperor Constantine moved his capital in 324 to Byzantium and renamed it Constantinople, left social disintegration in its wake. The population of Europe declined by half between 200 and 600, and Europe experienced its "Dark Age" of migrations, warfare, and breakdown of urbanization.

In Constantinople the remnant of Greco-Roman civilization developed into the Byzantine Empire. All pagan ceremonies were banned in 392, and the pagan temples of antiquity were destroyed; Christian rulers with complete political and religious authority refused to tolerate other religions. Eventually, in 1054, a formal schism developed between the Latin Church in Rome and the Orthodox Church in Constantinople. The Byzantine Empire survived until 1453 but gradually lost power, so that by the end of the twelfth century, two-thirds of the Christians in the former Byzantine Empire had become Muslims.

Before it ended, the Byzantine Empire had a decisive influence on the emergence of Kievan Russia, Vikings, known as Varangians in Russia, dominated the two large rivers in Ukraine-Russia—the Dnieper and the Volga rivers. The Varangian elites lived in cities—Kiev on the Dnieper and Novgorod on the Volga—while the local Slavs farmed for them. In 989 Vladimir I, ruler of Kievan Russia, chose Orthodox Christianity rather than Islam for his people because, it is said, of the magnificence of the city of Constantinople and its cathedrals, and because he believed Russians could not do without their vodka.[2]

Farther east, in Iraq and Iran, the Parthians ruled (247 BCE–224 CE), followed by the Sassanian Empire (224–651 CE). The armies of these rulers fought with large horses, whose size could only be sustained by hay made from alfalfa, which required lots of water from irrigation. During this period the cities of Mesopotamia flourished with considerable prosperity and cultural creativity. Sassanid farmers adopted crops from India and China—cotton, sugarcane, rice, citrus fruits, and eggplant. The Sassanid rulers established the Zoroastrian faith, as Constantine had established Christianity; both sets of rulers used religion as an instrument of politics and both practiced religious intolerance.

In India the Gupta dynasty from about 350 to 535 CE provided the political stability for the classic age of Indian culture. Its prosperity rested on intensified agriculture, particularly on rice introduced from southeast Asia into the western areas of India, which required cutting down forests. Cinnamon, pepper, and cotton textiles were widely traded on the roads to China, around the Mediterranean Sea and the Indian Ocean. The cotton plant may have been native to India; it matched the environment well, and Indians excelled in all aspects of its preparation for cloth.

Buddhist and Hindu holy men traveled the trade routes within India and across central Asia to China, bringing the hope of salvation to millions of people. Hinduism achieved some doctrinal definition in this period, while Buddhist monks and laypeople developed rituals appropriate for their distinctive roles. Buddhist monasteries, supported by admiring laypeople, became as important an institution as the portable congregations of the Judeo-Christian and Islamic traditions. All of these religions helped ease entry into the network of empire and trade for millions of ordinary people.

As discussed in the last chapter, the trade routes across Eurasia were called the Silk Road. Never a single road, these routes consisted of branches and segments, more or less significant in various times. The heaviest use of the Silk Road took place during the first millennium of the Common Era— until the eleventh century, when sea routes expanded and became more significant. During the first millennium, a caravan from Changan, China, took about four months to reach Samarkand and Bokhara (then Sogdiana, now Uzbekistan and Tajikistan), a journey of 2,500 miles (4,000 kilometers) over unsettled deserts, mountains, and grassland (Fig. 7.1).

The silk that constituted China's chief export remained a mystery fabric to Greeks and Romans for many years. They heard many possible explanations, such as that it was made from bark on trees. Not until the mid–sixth century did the Byzantine emperor learn from two monks that the cloth was a product of silk worms feeding on mulberry leaves.[3]

By the first century CE silk clothes were popular on the streets of Rome among its wealthy citizens. Much consumption of silk, at both ends of the Silk Road, was devoted to religious activities. Christian priests used purple silk embroidered with gold silk thread for their vestments. Kings, priests, and saints were shrouded in silks at their burials; even burials from long ago were dug up and shrouded in silk. In the Buddhist areas, yards of silk were used for banners, sometimes tens of thousands at one monastery. Buddhist laypeople made donations of silk to monasteries as a reward for the monks' intercessions and as a way to gain merits for future life. The monks, in turn, traded silk for daily provisions and for the "seven treasures" used to decorate their *stupas*, or shrines: gold, silver, lapis lazuli, red coral, crystal, pearls, and agate. During affluent times, Buddhist monasteries thus became significant economic entities.[4]

In addition to silk from China, other products moving along the Silk Road included: large horses, furs and woolen carpets from central Asia, cotton textiles, pearls and crystal from India, coral from the Mediterranean, glassware from India and the Roman world, and fragrances and spices from India, the Arabian Peninsula, and Africa.

Two accounts of travel along the Silk Road exist to tell us what the journey was like. Both were written by Chinese Buddhist pilgrims, traveling from China to India—one by Faxian (fah-shee-en), who died between 418 and 423, and the other by Xuanzang (shoo-wen-zahng), who died in 664. On their journeys both men stayed along the way in Buddhist communities and monasteries that previous generations had established.

Islam Arises and China Recovers (600–1000)

The second half of the first millennium of the Common Era saw the meteoric rise of Islam, starting in 630 in the Arabian Peninsula and spreading across the arid areas of the central core of Afro-Eurasia from the Pyrenees to the Indus River by 750.

People in the Arabian Peninsula at that time specialized in camel trade, carrying goods north from coastal cities in Yemen on the Arabian Sea to settled farming lands in Palestine, Jordan, and Syria. Polytheism was still the native religion, focused on natural forces and celestial bodies. Travelers carried the ideas of Christianity and Judaism, which became familiar in the peninsula. Some sources indicate a monotheistic cult existed, heavily influenced by Jewish practice and thought.[5]

The prophet of Islam, Muhammad, was born in 570 in Mecca, a caravan city near the Red Sea coast halfway between Yemen and Syria. An orphan reared by his uncle, Muhammad engaged in trade as a merchant and married a wealthy widow, Khadija, after successfully leading several of her caravans. In about 610 Muhammad began meditating in the mountains near Mecca. During one vigil, the "Night of Power and Excellence," a being that Muhammad understood to be the angel Gabriel (Jibra'il in Arabic) spoke to him. These revelations continued until his death; he recited them in rhythmic and rhymed verse in public but never wrote them down.

Muhammad's earliest revelations called on people to submit to the one God, Allah, who had created the universe and everything in it. At the end of time, people would be judged and the blameless would go to Paradise to enjoy the delights of the flesh, while the sinful would taste hellfire. Muhammad revealed himself as the last of Allah's messengers in the tradition of the Jewish prophets and Jesus, calling on each person to become a Muslim, or one who makes a submission, *Islam*, to the will of God. People who submitted would constitute the *umma* (community) uniting all Muslims in a universal community of social equality.

Since the leaders of Mecca did not accept Muhammad as the sole agent of the one true God and persecuted his weakest followers, Muhammad and his followers fled in 622 to a nearby city, Medina, 215 miles (340 kilometers) to the north. There he led a large community, conducting sporadic war with Mecca, which surrendered in 630. Muslims start their calendar with the year of the flight to Medina, known as the *hegira*; the Roman year 622 is year one for Muslims.

In 632 Muhammad died after a brief illness, leaving no arrangements for succession to his leadership. The father of one of his wives, Abu Bakr, became caliph, the spiritual and political leader; he had Muhammad's revelations written down into the Quran, which assumed its final form about 650. Conflicts over who should be caliph ensued in 656 and again in 680 and resulted in a permanent split between Shiite and Sunni Muslims.

Before Muhammad died, Muslims under his leadership were able to unite most of southern and western Arabia. After his death, Muslims united the whole Arabian Peninsula and won decisive victories over the Byzantine Empire and over the Sassanians in Persia (now Iran) between 634 and 651. The only material basis for these victories was camels— Muslims could supply armies across the desert landscape. Possibly the Muslims' conviction that Allah was on their side mattered most to the outcome of their battles. Other possible explanations for the rapid Muslim expansion include overpopulation of the Arabian Peninsula, whose people had begun migrating even earlier.

Muslims set up permanent military camps across a wide area—two in Iraq, one in Egypt, one in Tunisia. They moved their capital to Damascus in 661. Twice they failed to conquer the city of Constantinople, but they succeeded across northern Africa, and in 710 they crossed the Strait of Gibraltar into Spain. They pushed across Spain to the Pyrenees and into France, where a raiding party was defeated in 732 by French nobles at Tours 150 miles (240 kilometers) from the English Channel. France probably was not in real danger of being conquered and absorbed, but Muslims had struck terror into the heartland of Europe.

The decisive battle for Christian civilization took place at Constantinople in 717 to 718; if the Byzantine Empire had not held, despite tremendous losses, Europe might have become Muslim. The Muslim conquests of the seventh to the ninth centuries proved culturally decisive, however;

Christians eventually recaptured only the Mediterranean islands and Spain. In a bit more than a century after Muhammad's death, Islam went from the faith of one merchant in Mecca to an imperial state stretching from the Pyrenees Mountains in northwest Spain to the Hindu Kush—one of the most dramatic expansions in world history. Muslims did not force religious conversion and, except for Egypt, expanded into areas with sparse populations.

By 724 Muslims had reached the western frontier of China. They introduced gold and silver coins, called *dinars* and *dirhems*, inscribed in Arabic religious phrases, which circulated as one monetary exchange from Morocco to the edge of China. A consistent system of laws and contracts also was in place, facilitating trade.

In 747 a new family, the Abbasids, took charge of the caliphate and moved the capital from Damascus to Baghdad, where they ruled for over 500 years until Mongols killed the family in 1258. Two centuries of glory for Baghdad began with the Abbasids. The splendor of the court is reflected in the stories of the *Arabian Nights*, set in the time of the poetry-loving caliph Harun al Rashid, who ruled from 776 to 809. Cultural currents from Greece, Iran, central Asia, and Africa met in the capital, giving rise to rich literature, facilitated by the introduction of papermaking from China just as the city of Baghdad was being built in the 760s.

Devoted to books, Muslims produced more in their centuries of ascendancy than did any, and perhaps all, previous cultures. Papyrus had been harvested to near extinction, and in the middle of the second century unknown people in an unknown place invented a new form of book, the codex. The codex form consisted of leaves of parchment or papyrus, later paper, folded and bound between covers, the same as in present-day books. Muslims learned papermaking from Chinese prisoners of war, and by the tenth century it had largely displaced papyrus for writing in the Muslim world. Muslims also introduced the use of linen rags, beaten to a pulp, for papermaking in place of the mulberry bark used by the Chinese but unavailable in the Islamic world.[6]

Muslims called their Iberian territories al-Andalus. (Iberia was the ancient Latin name for the Spanish and Portuguese peninsula.) There they introduced the most sophisticated agricultural economy in Europe, featuring new crops of citrus fruit and sugar, and new irrigation systems. Cordoba, Seville, and Toledo grew larger than other cities in Europe. Cordoba and Grenada became centers of learning in which germinated the intellectual revolution of the eleventh and twelfth centuries of Western Europe (Fig. 8.1).

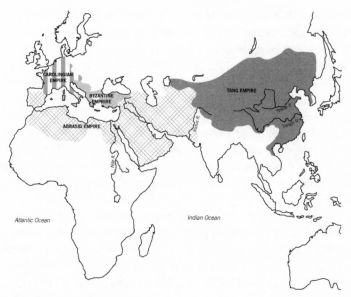

8.1   Ninth Century Afro-Eurasia

On the eastern side of Afro-Eurasia, after the fall of the Han dynasty, the Chinese people experienced more than 300 years of disorder and instability. During this time Buddhist and Taoist thought rose precipitously and became dominant in some areas of China.[7]

By the end of the sixth century the Sui family (589–618) succeeded in uniting China again and completing a new system of canals, accomplished by 5.5 million people laboring under 50,000 policemen. Under the Tang dynasty (618–907), China became the most advanced society in the world. With the Muslim empire providing stability from the Hindu Kush westward, and with Turkish people, mostly Uighurs, controlling the steppe lands, trade and travel could resume at high levels of participation as low-risk activities.[8]

Under the Tang, China consolidated control of its southern coastal area, increasing its access to the Indian Ocean. Chinese seamen excelled in the design of large seagoing ocean vessels; their ships carried two times as much volume as those from Constantinople or from Baghdad. Exports from China consisted mainly of superior silks and porcelain, a special kind of clay developed under the Tang. Chinese exports dwarfed other world trade. Anecdotes claimed that Chinese ships outnumbered others by a hundred to one; they did carry twice as much.[9]

Imports changed China. Men shifted from wearing robes to the pants favored by horse-riding Turks in central Asia. As the Chinese learned how to produce cotton, it replaced hemp as the most commonly worn fabric. Grape wine, tea, sugar, and spices modified the Chinese diet. All this trade was conducted with no central banking system; the Tang were wary of great accumulations of wealth. Individual scholars and landholders loaned money at interest.

The Tang capital at Changan became the largest urban center in the world with nearly 2 million people, 1 million within the walled area of thirty square miles. China had twenty-six cities with a population over half a million. Twenty percent of its citizens lived in cities, the most urbanized society of its day.

Under its Tang rulers China welcomed ideas and culture from outside its borders. Many cities had special areas set aside for traders and merchants from abroad who came to live in China to conduct business for their landsmen. Many of China's leading figures were of foreign origin; for example, the general An Lu-shan was Sogdian and the poet Li Po was born in Afghanistan.[10]

The Tang dynasty relied on paper to conduct its bureaucracy. A court eunuch, Tsai Lun, had invented paper production in the early second century of the Common Era. Printing experiments began in China in the sixth century. Under the Tang, literacy rates were 15 to 20 percent compared to at most 10 percent in Europe at that time. Paper money was first issued in the eleventh or twelfth century. Paper was also used to cover windows, as glass production required the burning of too much wood by Chinese standards.[11]

In about 850 Chinese alchemists seeking the elixir of immortality stumbled onto gunpowder, made from saltpeter (potassium nitrate), sulfur, and carbon. By the late twelfth century, when gunpowder came to the attention of the West, the Chinese had developed it through many stages and had perfected the barrel gun and the cannon.[12]

For over a century the Tang dynasty controlled much of central Asia. Its westward expansion ended in 751 in the significant battle of the Talas River, which took place in the area between Tashkent and Lake Balkhash, in what is now Kazakhstan. Arabs, Turks, and Tibetans defeated the Chinese, and from this time Tang power began to decline, yet the battle also stopped Islam's eastward expansion.

After the battle of the Talas River, knowledge of the technique for making paper was passed to Arabs by Chinese prisoners of war. Muslims established a paper mill in Baghdad; paper became widely used in Egypt by 1000 and spread from Muslim Spain to the French Pyrenees, where the first European paper mill was built in 1157.[13]

Four years after the battle of the Talas River, the huge, overweight, diabetic General An Lu-shan led his province, Hopei, in revolt against the lavish, cosmopolitan court in Changan. Chinese military governors suppressed the rebellion, then seized power themselves. Heavy tribute to steppe leaders ensued. By the mid-800s the Tang dynasty was experiencing political and military decay and the decline of foreign trade.

As the prosperity of the empire declined, a cultural backlash set in. Confucian advisors to the emperor persuaded him that their woes could be traced to foreign influences, particularly to Buddhists, whose monasticism was undermining the Chinese family and the tax base of the dynasty (monasteries paid no taxes). In 845 The Chinese government tried to purge the country of Buddhist ideas by crushing the monasteries, secularizing 260,000 monks and nuns and destroying temples and shrines. Buddhist ideas survived, since many had been absorbed into Confucian practices, and monasteries became legal again in later times, but the cosmopolitanism of the Tang court dissolved under the reestablishment of orthodoxies and did not reappear for centuries. Smaller states succeeded the Tang dynasty after 907.[14]

The Edges and Limits of the Afro-Eurasian Network

So far we have examined urban life and trade in the core areas connecting the Afro-Eurasian network. Now we shift our perspective to that of the peripheral areas, where people experienced urbanization, long-distance trade, and universal religions coming at them as new, disturbing experiences.

People in cities depended for the most part on food brought in from outlying agricultural areas. Over much of Eurasia another pattern of society existed, that of nomadic pastoralism, in which people depended on certain animals and followed them around to new pastures as they needed fresh supplies of grass. In the heartland of northern Eurasia, people depended on horses, which they had tamed first in what is now southern Ukraine, by about 4000 BCE. By about 1500 BCE the people of the steppes had developed a culture dependent on mounted horses.[15]

Nomads had special incentives for trading, since their own products were limited by their need to keep moving. They could survive on their animals alone, but they greatly desired grain, fabric, and metals. Their options for acquiring these products were trading or raiding. Since they were superb warriors, they could negotiate payments in return for protection of local farmers, as they often did with Chinese rulers and governors. From 386 to 534 the Toba confederation of nomads even ruled the civilized populations in northern China.

On the western side of the steppe lands, a succession of new steppe people arrived in Eastern Europe during the years 200 to 1000, driven west by nomads further east. The new groups were called Huns, Avars, Bulgars, Khazars, Pechenegs, Ostrogoths, and Magyars. The Huns (374–453), who were headquartered on the Hungarian plains, raided far and wide into Gaul and the Rhineland, withdrawing from invading Italy ostensibly at the plea of Pope Leo the Great (440–461). When the Huns' leader, Attila, died in 453 and the plague struck the rest of them, the Huns dispersed.

The Huns were not able to conquer the Byzantine Empire. It held at the Danube River, but Gaul, the British Isles, the Iberian peninsula, and northern Africa fell to Germanic peoples fleeing the Huns, and the legacy of Rome in the west went underground. Over the next 600 years, culture wars played out as Roman and Christian traditions competed with those of Germanic peoples and those of succeeding waves of invaders—more Germanic invasions from 568 to 650, followed by invasions from Vikings, Magyars from Hungary, and Muslims from Spain. During this period western Europe reverted to a fringe area of the Afro-Eurasian network, lacking central government and secure trading routes.

In the matter of language, the vulgar Latin of the Roman tradition evolved rapidly into the Romance dialects—Portuguese, Spanish, French, and Italian—which held their own against the languages of the new invaders, except in the northern areas, where Latin gave way to Germanic and Scandinavian languages. The Germanic tribes of Angles and Saxons conquered England between 410 and 442, creating a Germanic-based language and bringing the soil under cultivation for the first time.

In Italy a pious hermit, Benedict of Nursia, organized several monasteries of monks, each headed by an abbot. Benedict wrote the manual for monastic life in western Europe, called the Rule of Benedict, which emphasized poverty, celibacy, and obedience to the abbot. Without the work of monks over the next centuries, most of the surviving ancient Latin works would likely have been lost. Only one copy of nine complete plays by Euripides, one copy of Tacitus, and one copy of *Beowulf* made it through the dark times.[16]

From the south, Muslims captured Spain from the Visigoths in 711 and tried to conquer Gaul, as described earlier. The major threat to Europe, however, came from the Viking population in Norway, Denmark, and Sweden, commonly called Norsemen by Europeans and Varangians by Russians and Ukrainians. The Vikings had developed a society based on three classes: slaves, free peasants, and warrior-chieftains. Without heavy plows until about 1000, Vikings consumed mostly oats and barley, sheep and goats, cattle and fish. They traded in a far-flung network, shipping down the Dnieper and Volga rivers to the Black and Caspian Seas and beyond to Baghdad on the Tigris River. They captured slaves from the British Isles and from Slavonic areas to trade to Europeans and to Muslims. Fur was their other chief commodity—bearskins, sable, marten, and squirrel. They also traded lumber, reindeer hide, salt, glass, horses and cattle, white bears, falcons, walrus ivory, seal oil, honey, wax, woolens, and amber (Fig. 8.2).

The Vikings believed that a tree of fate, the Yggdrasill, occupied the center of the divine world, where the gods sat in council every day. The tree had three roots, one to the world of death (Hel), one to the world of frost giants, and one to the world of people. The Yggdrasill, they believed, held up the universe; they considered a special tree in their town of Uppsala the earthly replica of Yggdrasill. Every nine years all the Viking people gathered at the temple in Uppsala to make sacrifices to their gods for nine days. On each day a sacrifice was made of nine living creatures, including one man, and the bodies were hung on trees near the temple.

What the Vikings believed is known because they created an alphabet called runes, which appeared at the end of the second century or the beginning of the third. The letters were made of straight strokes, suitable for carving into wood or stone. The Vikings adopted about two-thirds of their

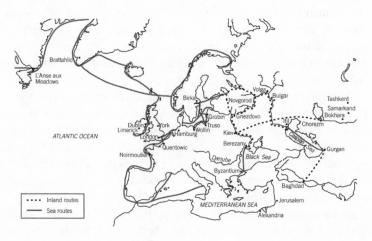

8.2   Viking Trade Routes

alphabet from Latin and Celtic letters, inventing the remainder. Sagas of their travels to Iceland remain to tell us how they viewed life.

The brief period of Viking expansion started shortly before 800 and ended by 1070. Overpopulation is conjectured as the chief cause for this swarm to neighboring lands, as men sought land, wealth, and fame in places with no central empire to protect the inhabitants. Possibly the Vikings were responding to the conquest of the Saxons by the French king, Charlemagne, which brought him to the border of Denmark. Improved techniques of shipbuilding contributed to the expansion. Vikings learned to make ships of oak, up to seventy feet long, which drew only three and a half feet and carried from thirty to over a hundred men. They made masts of pine and sails of wool. Few human artifacts match the beauty of a Viking ship.

By 820 Vikings had settled Novgorod. In 839 they invaded Ireland, and from 866 to 878 they invaded England, where they took five eastern districts to rule. They attacked the coast of Normandy from 841 to 884 and were given land in the Seine basin in exchange for protecting Paris. Later they sailed farther south in France and around the coast of Spain until they reached Barcelona, Marseilles, and the coast of Italy. They sailed west to set up colonies in Iceland (875), in Greenland (982), and briefly in Newfoundland soon after 1000. Remains of Viking-built houses were discovered at L'Anse-aux-Meadows in Newfoundland in 1962, proving that the Vikings had been there.

In about 965 Harald Bluetooth (940–985), the king of Denmark, converted to Christianity and recognized Otto I, the Holy Roman Emperor of the German confederacy, as his overlord. Norwegians were forcibly converted in about 995, although the temple at Uppsala was not destroyed until well into the twelfth century. Viking influence, brief though it was, proved a significant contribution to the making of western European culture.[17]

After the fall of centralized government in Rome, farmers in western Europe, under attack from nomads on all sides, took shelter with local landowners strong enough to offer hope of protection. The Germanic tradition of manors, or self-sufficient farming units, became widespread. Agricultural workers, known as serfs, came to belong to the manor and could not leave, in exchange for protection from landowners who, under conditions of nearly constant warfare, became noble knights. Outright slavery, a mainstay of the Roman economy, diminished. Cities lost population, roads fell into disrepair, trade languished, literacy declined—no won-

der Europeans call this period medieval or the Middle Ages, coming between the grander Greco-Roman civilization and the Renaissance of the
fourteenth century.

Unexpected though it may seem, agricultural production during this
period increased due to improvements in the plow. In northern Europe
heavy clay soils made plowing more difficult than in other kinds of soil.
Heavier plows were developed, eventually ones so heavy that they required
six to eight oxen to pull them, with moldboards that turned the dug-up soil
to one side. By 1000 carpets of grain fields (wheat, barley, and rye) stretched
across the landscape of northern Europe, where the basic diet consisted of
beer, lard or butter, bread from wheat, barley or rye, pork from forest pigs
living on acorns, and wild game. In southern Europe the human diet was
based on wheat, wine, and olive oil.

The displacement of Roman governance by Germanic states across
western Europe could have resulted in the disappearance of Christianity as
an important cultural force. Yet it survived to dominate Western civilization; how come? [18]

The papacy in Rome did not collapse. It offered a strong source of
unity, authority, and organization, especially beginning in the sixth century
when popes proved willing to ally themselves with Germanic rulers, who
converted early, and to absorb enough Germanic traditions to make Christianity palatable to polytheists. Not that it was easy to wipe out pagan practices; clerics in some parts of western Europe were still issuing prohibitions
against the worship of trees, rivers, and mountains as late as the eleventh
century.

The events that led to the domination of Christianity took place primarily in the British Isles and in France. In Britain the Irish Celts, never
conquered by Rome, had been converted to Christianity by St. Patrick, a
Romanized Christian Briton, in the first half of the fifth century. After the
Saxons and Angles took over England, Christianity survived only in Ireland, where monasteries sent monks and scholars as missionaries who promoted "Celtic Christianity" in England. Pope Leo I (590–604) also sent
missionaries from Rome. In 664 Christians in England met and decided to
choose the Roman over the Celtic church. During the eighth to tenth centuries the Saxons in England fought invading Vikings, who gained posses-
sion of a large territory in eastern England, as noted earlier. A Danish king,
Canute, ruled England from 1017 to 1035, and in 1066 a Norman, William
the Conqueror, descendant of Vikings who had settled in Normandy, seized
the throne. William agreed to pay the customary dues to the Church of

Rome but refused to acknowledge the pope's authority over him; the English church recognized no new pope and accepted no papal commands without the king's assent. Hence, the English king chose Christianity but on his own terms.

In what is now France, a Germanic group known as the Franks originated in the Rhineland. Some of them had lived in the Roman Empire and had converted to Christianity. After the dissolution of the Roman empire, the Franks extended their kingdom southward all through what is modern France. Their first great leader, Clovis, who ruled from 481 to 511, converted to Roman Christianity in 508, apparently through his wife, Chrodechildis, or Clotilde in French, who was already Christian. The Frankish kings made alliances with the popes, gave them land taken from defeated Lombardians, and, under Charlemagne (r. 768–814), pushed eastward into Saxon territory to the Elbe River and along the Danube below Vienna. In 772 Charlemagne and his troops destroyed a grove of trees sacred to the Saxons, whose religion closely resembled that of the Vikings.

Since the Saxons strongly resisted giving up their religious practices, Charlemagne forced Christian baptism on them, imposing the death penalty for those who violated the Lenten fast, killed a bishop or priest, cremated the dead in Saxon fashion, refused baptism, plotted against Christians, or disobeyed the Frankish king. Saxons fought for over thirty years to resist Charlemagne and Christianization, but monks and priests followed the military victories of the Franks, establishing a secure enough presence to Christianize Germany over the long term.[19]

Charlemagne's empire in the eastern lands, in what is now Germany, fell into local duchies until a strong monarch emerged, Otto I, who could win papal coronation as the Holy Roman Emperor in 962. The Holy Roman Empire continued as a loose confederation of German princes who named one of their own to the office of emperor.

By 1000 large landowners in Europe had become knights in armor defending their manors and allying themselves with local kings, who had become Christians over the centuries since Roman rule had vanished. Christianity itself had shifted in the translation from a pacifist religion, taught by the rabbi Jesus to marginal social groups in an isolated province of Rome, to a warrior religion used to defend the people of Europe and the Byzantine Empire.[20]

While western Europe experienced its dark age of migrations and warfare, Africa was coming more into the network of trade and stability emanating from the Muslim world. North Africa converted to Islam, as described earlier, while sub-Saharan Africa continued to be something of a special case, isolated from the main currents of world trade far longer than most of the Eurasian landmass (Fig. 8.3).

The use of camels as pack animals came to Africa from Arabia sometime before the start of the Common Era, reaching Lake Chad at around 300, when caravan traffic across the Sahara began. Since camels are superior to other pack animals, carrying more for less food and water, the Sahara became a favored region of cheap transportation. Egypt and northern Africa came under Muslim control between 636 and 711, after which African contacts with the rest of the world were chiefly through Muslims. During the early centuries of trade with Muslims, West African rulers resisted accepting the Muslim faith, with its literacy and world participation, because it required repudiation of local religious traditions that gave kings their claim to sacred powers. The first known African king to convert did so in 985.[21]

Along the eastern coast, sailors from what is now Indonesia reached Madagascar in about 500, bringing bananas, yams, and taro root, which the native Bantus cultivated, allowing settlement of more forest zones. Islam spread along the coast, where a common culture and language developed, based on African grammar and vocabulary but enriched with many Arabic and Persians terms and written in Arabic script. In time, the people and the language became known as Swahili, from the Arabic name *sawahil al sudan,* meaning "shores of blacks." The Islamic traders did not go inland, where Bantu-speaking Africans remained content with their own religions for centuries to come.

Camels that could cross the Sahara were useless in the sub-Saharan humid climate because of the tsetse fly and the trypanosomes it carried, which caused sleeping sickness. Other diseases in sub-Saharan Africa—malaria and yellow fever—proved fatal to people not used to that disease environment. Geography also kept people out of sub-Saharan Africa; the largest rivers, the Niger and the Congo, were cut off from the sea by rapids or waterfalls near their mouths. A complex array of parasites kept the local population in check. The interior of Africa remained isolated from the rest of the world until the nineteenth century, without the density to develop urban societies; local traditions, using about 2,000 languages, remained characteristic of sub-Saharan Africa.[22]

Africans developed iron smelting, beginning sometime early in the first millennium CE, possibly by themselves or by importing the idea and/or technique. The Bantus, who lived on the edge of the rain forest near known sites of early iron smelting, close to the modern boundary of Nigeria and

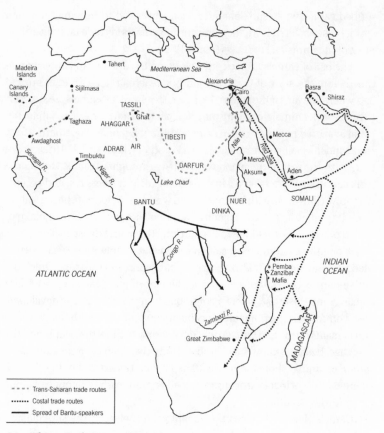

8.3　African Trade Routes About 1000 CE

Cameroon, moved south and took iron smelting with them into southern Africa by 800.

In the hot, wet tropical forests and humid savannas of sub-Saharan Africa, all goods that were traded had to be carried on the heads of people. No other pack animals could survive. This fact favored the lightest, most valuable commodities, particularly gold. Two kingdoms based on trading gold emerged before 1000—Ghana in West Africa and Zimbabwe in the east.

Ghana appears in an Arabic text of the late eighth century as a "land of gold." Covering parts of Mali, Senegal, and Mauritania, the kingdom of Ghana was developed by the Soninke people, who traded gold dust to the Berbers along the northern coast in exchange for copper and manufactured goods. The ancient capital of Ghana was a double town before 1000, with one area for merchants of all origins and another area for military and political leaders and their followers.[23]

On a plateau south of the Zambezi River, another powerful state emerged, based on gold mined locally and carried to the coast. The capital of this state, now known as the Great Zimbabwe, had about 18,000 inhabitants at its height in about 1400. Historians suspect that people in the capital used up nearby forests for firewood, while their cattle overgrazed nearby pastures, hastening the state's decline in the fifteenth century. On the whole, however, the shorelines of Africa remained a frontier to the Afro-Eurasian network before 1000, and the interior remained a frontier long after that.

On other frontiers, the Vikings were settling Iceland, Greenland, and Newfoundland, as mentioned earlier. A lot of seafaring was going on in the Indian and Pacific oceans, but nothing written, like the Icelandic sagas, remains to tell the story. Polynesians settled the Easter Islands and Hawaii in about 400 and New Zealand in about 1300, but they had no contact with the rest of the world. By about 400 the western Pacific and Indian oceans had become one large sea room. Ships, able to navigate long distances, no longer had to hug the coast or pay local duty. Many goods circulated widely—pepper and cotton from India, porcelain and silk from China, nutmeg and cloves from Indonesia, gold and ivory from Africa. The tempo of complex interactions had increased dramatically in eight hundred years.

In summary, the 800 years between 200 and 1000 witnessed a dramatic intensification and expansion of the Afro-Eurasian network. Ships and caravans strengthened and extended trade and communication. Before 200 agrarian civilizations existed only spottily across the area of the core network; by 1000 parts of Africa, the whole of southeastern Asia and the south Pacific islands, Korea, Japan, northern Europe, and the steppes had been added to the network, which by then included some 200 million of the world's 253 million people.[24]

This period was characterized by the spread of religions of salvation into areas where people had formerly worshipped local gods or natural spirits. Islam culminated the transition from an identity based on local ethnicity and localism to one based on a universal religion. When Muslims encountered polytheistic peoples in northern Africa and among the Turks of the western steppes, mass conversions followed. Christians met with success among Celtic, Germanic, and Slavic peoples. Vikings resisted until 1000, but by then all Europe was Christian except pockets along the southern shore of the Baltic, where people were the last to convert in 1387.

Buddhist ideas spread into central Asia, China, and southeast Asia. The commonality in the religions of salvation lay in their directing human aspirations toward an external, transcendental world—heaven, paradise, nirvana, union with Shiva and Krishna. They gave hope of redress in a future life to people distressed by the realities of urban life. Cities were unstable, inequitable in times of prosperity, and liable to break down in times of retrenchment. The new religions sustained hope and helped maintain the social differentiation required by city organization. The increase in urban societies was connected to the increase in conversions to salvation religions; both were a common feature of human life in Afro-Eurasia from 200 to 1000 even as many nonurbanized areas remained.

Costs of Complexity

There were costs associated with this increased complexity. To achieve it, humans had to divert more of Earth's energy to flow through their systems. One way they did this was to cut down trees.

If one could have hovered above Eurasia in a spacecraft, once in about 200 and again in about 1000, the most notable difference would be the vanishing forests. Across China all the way to India, one of the world's great blocks of tropical forest disappeared as trees were removed for rice paddies to feed more people. Across Europe virgin forests were felled to plow the heavy soil for grain crops to feed more people. People who worshipped the forests converted or were converted into believing religions that promised salvation later, in exchange for more human survival and complexity now.

Despite these losses of forest, more than 75 percent of the Earth's trees remained. As currently estimated, deforestation did not reach 25 percent of the trees that remained in 1985 until about 1700, with 50 percent reached by about 1850 and 75 percent by about 1915. Even though the pattern of deforestation was evident by 1000, its pace had not reached the rapid acceleration of the past three centuries.[25]

The shift in the balance of power between humans and all other animal species also continued. Humans increased their dominance over all other mammals, hunting them into small remnants and pressing many of them into the service of human needs as food and pack animals.

The total number of people on Earth decreased slightly from 200 to 1000, from 257 million to about 253 million people. In 1000, Europe, excluding Russia, had about 30 million, 14 million less than it had in 200. China had about 56 million people, or about 22 percent of the world's population,

Africa had about 15 percent of the population, and the Americas had about 7 percent.[26]

By 1000 five patterns common to the history of the human species could be discerned in Afro-Eurasia, at least in retrospect. People were gradually, but over time consistently, increasing in numbers. As their total numbers increased, so did the concentrations in which they lived. As their concentrations increased, so did the degree of stratification, organization, and specialization in skill and knowledge. In any given year over the past ten millennia, these five aspects of human living could be found in greater size and degree than before. The increase was not steady; it fluctuated. But charted by thousand-year intervals, the patterns held, with the prevailing pattern of increasing complexity.[27]

The last three chapters have told the human story in Africa and Eurasia. Now it is time to turn the globe on its axis to bring into view the Americas, still not regularly connected to Afro-Eurasia but inhabited by humans for at least 13,000 years, and possibly for as long as 30,000 years, and where a separate experiment in the emergence of complex societies was being conducted.

Unanswered Questions

1. Was there contact between Africa and the Americas prior to 1492?

Some tantalizing evidence suggests there may have been. The currents in the Atlantic Ocean flow off the coast of Africa in two places—the Cape Verde Islands and the Senegambian coast—toward the northeast coast of South America and the Caribbean. The Norwegian adventurer, Thor Heyerdahl, had a ship of papyrus reed built by Africans to ancient specifications and succeeded in sailing in it from Safi to the Barbados Islands in 1969. Others duplicated this trip. Indeed, the Portuguese captain Pedro Alvares Cabral verified this current when his ships, on their way around Africa, were accidentally carried to the coast of Brazil in 1500.

Evidence of African attempts to reach America appear in a book by an Arabic geographer in Damascus, Ibn Fadl Alah al-Umari (1301–1349), who was in Cairo twelve years after the emperor of Mali, Mansa Kankan Musa, passed through in 1324. The man who hosted Mansa Musa told al-Umari stories recounted by Mansa Musa, including an account of how he got to be king. It seems that his predecessor, Mansa Muhammed, believed it was possible to reach the limit of the Atlantic Ocean. He provisioned 200 ships and sent them out from the Senegambian coast (then Mali). Finally, one returned to report that the others had encountered a river with a powerful current in the open sea. The others entered and were not seen again, so the one turned back. Mansa Muhammed then outfitted a larger expedition, left Mansa Musa as king, and, in 1311, departed himself, never to be seen again. It is possible that the current was the mouth of the Amazon and that these sailors landed in Central America.

The hypothesis that Africans did reach the Americas before 1492 has been defended primarily by Ivan van Sertima, an anthropologist from Guyana and professor of African Studies at Rutgers University. His books include *They Came Before Columbus* and *Early America Revisited*; most mainstream historians have not accepted his evidence.

9

Emergence of American Civilizations

(200–1450 CE)

The Americas were the last continents of Earth to be inhabited by *Homo sapiens*. People got to Australia about 40,000 years ago but did not make it into what is now Alaska until possibly as early as 35,000 years ago or as late as 13,000 years ago.

Meanwhile, plants and animals in the Americas enjoyed a protracted period of development without having to confront the hunting prowess of humans. During the Great Ice Age, between 75,000 and 10,000 years ago, animals could easily cross the land bridge between Asia and Alaska. For example, the horse, which originated in the Americas, crossed over to Asia, and lions, the mammal that has come the closest to the worldwide distribution of humans, crossed over from Asia to America. During these tens of thousands of years, North America was home to five times as many species of mammals as there are now. Four types of giant ground sloths roamed around, one twenty feet high weighing three tons. The mix of mammals must have been amazing—giant rodents, three species of camel, woolly mammoths, mastodons, long-horned bison, glyptodonts (armadillo-like), huge lions and tapirs, cheetahs, saber-toothed cats, elephants, wolves, giant short-faced bears, and horses.

Between 16,000 and 10,000 years ago many of these animals became extinct. Why did all these extinctions happen in this short period of time? Experts cannot be certain; a complex interplay occurred between the warming of the climate and the arrival of people. The fact that many animals died can be explained either by their inability to cope with the rapid change in habitat or by the ease with which *Homo sapiens* could hunt them, or a combination of both.[1]

Humans on the Scene

For reasons still unclear, the immense ice sheets began to melt about 14,000 to 10,000 BCE. Profound changes occurred in the landscape as retreating glaciers left heaps of soil, rocky debris, and lakes dotting the Earth. Complex arrangements of organisms were fragmented and scattered, with a northward migration of plants and associated animal communities.

Onto this scene crept human hunters, possibly at the last possible moment before the rising seawaters engulfed Beringia to close off the land bridge. (Today the Bering Strait consists of about fifty miles of frigid water between Alaska and Siberia.) As a probable scenario, a small group of people came into Alaska in about 14,000 BCE; within a couple of thousand years a corridor opened in the glacier, enabling the Alaskans to come out to the Great Plains at about the site of modern Edmonton, Canada. At least two major waves of migration, possibly three, left from different parts of east Asia, dispersing swiftly across the two new continents, rapidly increasing their population as they traveled. By 9500 to 9000 BCE people had spread from the Canadian Plains to central Mexico, coast to coast. From the small initial group, let's say about a hundred, people apparently reached a population of about 1 million in 300 years. By 8500 BCE people had reached Tierra del Fuego at the southern tip of South America—8,000 miles in less than 1,000 years.[2]

What an Eden the Americas must have been for people after life in Siberia—huge animals easy to hunt, clouds of birds, a pristine, spectacular, warming wilderness. In the first 2,000 years of human habitation, an estimated 50 to 100 million animals disappeared, including many large animals that might have been domesticated.

The people who crossed the Bering land bridge arrived without pottery or domesticated animals, except possibly dogs. They remained hunters and gatherers until about 6000 BCE, when they began domesticating plants in four different areas: Mexico, the eastern woodlands of North America, the tropics of South America, and the Andean altiplano. Plant domestication happened simultaneously and apparently independently in these four areas, suggesting common climate shifts that favored plant growth. Taking all four zones together, Native Americans domesticated more than a hundred plants, among them some of the world's favorites: corn, potatoes, chile peppers, tomatoes, peanuts, tobacco, cacao, and coca. But even after agriculture began in the Americas, people remained primarily hunter-gatherers; 4,500 more years passed before permanent settlements emerged in Mexico and in the Andean zone.

The earliest crops domesticated in the Americas appear to have been chile peppers in Central America, sunflowers in the woodlands, pumpkins in New Mexico, and amaranth (a nutritious grain) in these three sites, and nuts in the Andes. Recent evidence suggests that large-scale tropical gardening may have taken place in Bolivia and along the rivers of Brazilian Amazonia by 2000 BCE; manioc may be as old as corn.[3]

In the Americas there were no wild wheat, barley, oats, or rice to domesticate. As discussed earlier, maize, or corn, the crop that would eventually support American civilizations, began as a wild grass with a tiny cob about the size of a human thumb. After a long process of genetic selection, domesticated corn emerged in about 5000 to 4000 BCE. Although corn yielded more calories per acre than did rice, wheat, or barley, it was slow to spread in the Americas because it had to adjust genetically to different day lengths in different climates before it could ripen with the seasons. The north-south axis of land in the Americas contrasted sharply with the east-west axis in Eurasia; it meant that crops could not spread at similar latitudes but had to adjust to the climatic conditions of different latitudes.

By about 3000 BCE Mexico had developed its distinctive cuisine of corn, beans, and squash. Since there were no animals to domesticate other than turkeys and dogs, Mesoamericans depended mostly on the combination of corn and beans for their protein. Mexican crops slowly spread into the southwestern United States by about 1200 BCE and into the eastern woodlands by about 1000 CE.

In South America two climates produced quite different foodstuffs. In the tropical lowlands, people domesticated manioc and sweet potatoes possibly by about 4000 BCE. Sweet potatoes somehow got introduced into the island chains of Polynesia, presumably from the Americas, suggesting some intermittent voyages long before European connections. Perhaps Polynesians, who reached Easter Island in about 400 CE, also touched American shores, taking sweet potatoes back home with them. No trace remains of such a voyage; the presence of sweet potatoes in Polynesia remains a mystery.[4]

Farther south, people in the high Andean mountains produced a distinctive culture and cuisine. Potatoes and quinoa, a lightweight grain of high protein content, were their staple carbohydrates. Potatoes were cultivated in 3,000 different varieties, from tiny purple ones to large white ones, an array that has been represented in North American markets in recent years. Andeans also had three kinds of animal protein—llama and alpaca, which they were able to domesticate, and vicuna, which they were not. (Together these three are referred to as camelids.) Llamas and alpacas were used as pack animals but were not milked or plowed. Andeans learned to freeze-dry potatoes (*chuña*) and animal flesh (*charqué*)—whence our word "jerky."[5]

Agricultural development went along slowly in the Americas. Without large domestic animals for plowing, fertilizing, and grazing, people could increase their food supply only slowly by labor-intensive irrigation of new land. Domesticated crops could not spread rapidly because of the north-south orientation of the latitudes. Not until about 1500 BCE did permanently settled communities emerge in Mesoamerica and in the Andes. A few grew rapidly in population, with increasing social differentiation, organized warfare, and monumental architecture—the signs of emerging agrarian civilization.

Urban Centers in Mesoamerica

The founding urban culture of the Americas, emerging in about 1000 BCE, belonged to the Olmecs, located on the coast of the Gulf of Mexico in the present states of Tabasco and Veracruz. This seems to have been a society of about 350,000 people living in small towns who built three ceremonial sites at which to gather intermittently. The Olmecs developed no system of writing with which to represent speech, but rather used glyphs, or pictographs, as mnemonic devices to remind them of certain gods or ideas.

Without writing to give us access to Olmec ideas, we can only make inferences from their material culture. They carved huge heads out basalt boulders, each with a unique, distinctive face and some of which stood 11 feet (3.4 meters) tall, weighed 20 tons, and had been transported by boat and/or over land some 50 miles. Their gods had dual male/female natures. The Olmecs loved jade, in which they frequently carved jaguars, or men transformed into jaguars.

The Olmecs were the first to play the "ballgame," or *tlachli* in their language of Nahuatl. The game used a long stone court (100–200 feet or 30.48–61 meters) shaped like a capital "I" with sloping sides. Players wore gloves, girdles, and deerskin hip guards. A rubber ball had to be hit with hips or knees, never with hands or feet, into hoops placed at both ends. Carvings on the stone walls of some courts depict the losing team's captain being decapitated—or was it the winning team's captain, since sacrifice was considered an honor?

The Olmecs carved a "Calendar Round," a complex calendar based on fifty-two years, with an overlap of two different day-count systems. One,

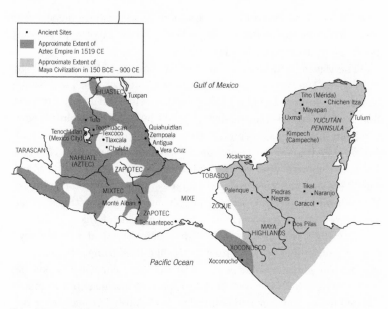

9.1   Mayan and Aztec Empires

the ritual system, consisted of 260 days divided into 13 months of 20 days. The other, the solar system, consisted of 365 days divided into 18 months of 20 days, with 5 extra, unfavorable, days at the end. Any day in the two systems would intersect every fifty-two years. With this cyclical view of history, the Olmecs apparently expected repetition of the essential patterns, such as royal dynasties and military invasions. They made a "long count" of years back to August 13, 3114 BCE, formulated on a base-20 math system, using a bar for 5, a dot for 1, and a shelllike character for 0.

The Olmecs began to decline as early as 400 BCE with some kind of collapse of the state structure. Was the cause an internal uprising, an external invasion, or crop failure? No one knows.

Before the collapse of the Olmecs, a rural provincial power of similar achievement, the Mayans, arose south of the Olmecs on the Yucatan Peninsula about 600 BCE (Fig. 9.1). There the climate and environment were harsher challenges—the heat made it difficult to keep land cleared, and extensive limestone deposits leeched nutrients from the soil. Slash-and-burn cultivation could not sustain a dense population; for that people had to use irrigation and raised-field systems of agriculture.

The raised-field system consisted of building up island fields by piling extra soil in swampy areas, leaving canals of standing water between them.

These raised fields, used by all Mesoamerican civilizations, appeared to float. Requiring extraordinary labor, they could be harvested by boat, a boon to people without pack animals.

The Mayans were never a centralized empire but rather, at their height from 600 to 800 CE, consisted of a group of about sixty small town-states sharing a culture and ideology. Their population reached at least 3 to 5 million, more than double the number of people living on the peninsula when the Spanish arrived six centuries after the collapse of the Mayans. The Mayan staples were corn, beans, squash, tomatoes, and peppers. They constructed elaborate water storage and delivery systems, but remained vulnerable to drought. Their rulers communicated with ancestors and gods through self-sacrificing or bloodletting, piercing their skin with cactus needles. Two women are known to have ruled Mayan kingdoms.

Only the Mayans in the western hemisphere developed a complete system of writing that could represent speech; it used some ideographic, or glyphic, elements and some phonetic ones to represent syllables. But they did not write everything phonetically, apparently because the ideographs seemed to have prestige and political and religious overtones. For books, the Mayans used bark paper coated with gesso and folded into screenlike books. So thorough were the Spanish in their destruction of Mayan culture that no book remains from prior to 900 CE and only four from after that. Mayans also carved their writing on stone monuments, walls, and funerary vases; it was deciphered only in the 1980s.

The Mayan state structure collapsed rather quickly, in a few generations, by about 900 CE, with widespread famine and the loss of half of its population. The reasons for the collapse are unknown, but recent studies have given a better idea of the factors involved. Ceremonial sites were abandoned quickly, with some carvings left incomplete. Warfare increased vastly; short-lived fortifications have been discovered. Ruling families and priests disappeared, apparently repudiated by commoners. The people who remained reverted to slash-and-burn agriculture with no provincial, only village, structure. The year 889, the end of a fifty-two year cycle, was commemorated at only three sites. It seems that political disorder accompanied the rapid loss of food, possibly caused by long-term, intense drought that started in about 750 and led to the abandonment of state structure in the Mayan lowlands after about 900. The tropical jungle reclaimed the ornamental temples, leaving only the mystery.[6]

Starting a little later than the Mayans, another Mexican civilization arose in the city of Teotihuacan, located thirty miles east of modern Mexico City. The rulers of Teotihuacan dominated the highlands and basins of south-central Mexico for more than eight centuries, from about 200 BCE until about 750 CE. In its heyday in the middle part of the first millennium, Teotihuacan was home to 100,000 to 200,000 people contained in about eight square miles. Obsidian, or volcanic glass, formed the basis of its wealth.

Culture in Teotihuacan perpetuated many characteristics derived from the Olmecs, including the calendar and the glyphic writing system. The pantheon of gods, too, was similar in character though different in name. The most prominent were the Flayed One, the Sun God, the Moon God, the Rain God (Tialoc), and the cult of the Feathered Serpent (Quetzal-coatl). The jaguar represented fertility of the Earth and the serpent the fertility of the sea. In the eighth century CE, desert people from the north successfully invaded Teotihuacan, and its people scattered into surrounding small communities.

The fall of Teotihuacan and, a hundred years or so later, of the Mayans, indicates that some sort of environmental crisis decreased agricultural production during this time throughout central Mexico. In the recovery, the organizing group proved to be the Toltecs, who founded the city of Tula. The Toltecs dominated for only 200 years, but they had a lasting influence, introducing metallurgy (copper tools and ornaments), bows for hunting, and a militaristic, oppressive way of life that included human sacrifice to placate the gods as a common occurrence. Their high priest, Topiltzin Quetzalcoatl, claimed to be the mortal form of their god of culture and civilization, called Quetzalcoatl, but Topiltzin was driven away by followers of the god of darkness and warfare, setting sail over the ocean and promising to return some day, according to legends. Following droughts, hunter-gatherers from the northern desert invaded, and the Toltecs fell in the early 1100s.[7]

Toltec culture was picked up and used by an obscure tribal group, the Mexicas, who, legends say, had wandered for 200 years in the provinces serving other groups. In about 1325 the Mexicas, later known as Aztecs, settled on an island in the middle of Lake Texcoco—one of a series of shallow lakes that covered 620 square miles on the plateau of the central highlands of Mexico 7,000 feet (2,100 meters) above sea level; it was surrounded by mountains and drained into the shallow lakes. Rich deposits of obsidian abounded. Rainfall was localized, variable, and unpredictable. In this environment the Mexicas/Aztecs created the densest concentration of population in pre-Hispanic America, possibly reaching 2 to 3 million, in their city of Tenochtitlan and its neighboring clusters.

On the lakes, which were more like swamps only three to seven feet deep, the Aztecs built raised fields of higher land, called *chinampas*, which could be drained and farmed. Families lived on these islands and watered them with buckets of water splashed from canoes floating in the canals surrounding them. The Aztecs constructed elaborate terraces and irrigation systems on the floor of the valley and up the slopes; the higher elevations had earlier provided forest products and wild game, but by the time of the Aztecs little wild meat was left, while only turkeys and dogs had been domesticated. Corn was the staple crop, supplemented by beans, squash, tomatoes, chile peppers, amaranth, and agave, which were fermented to make pulque.

On their island swamps the Aztecs built the city of Tenochtitlan, now buried deep beneath modern Mexico City. Tenochtitlan covered 5 to 6 square miles (13–15 square kilometers); at the center a walled plaza 500 yards square (48 meters squared) accommodated 8,600 men dancing in a circle. A twin city, Tlatelolco, hosted the largest market in Aztec times, rivaling the markets of Rome and Constantinople. Since there were no wheeled vehicles or draft animals, trade was dominated by lightweight goods—gold, jewels, feathers, cacao beans, and skins.

In 1428, a hundred years after their first settlement, the Aztecs defeated their overlord city (Atzcapotzalco) and organized an alliance with two neighboring cities to begin acquiring their empire. By 1519 they had 38 dependent provinces, which sent a annual tribute to Tenochtitlan of 7,000 tons of corn, 4,000 tons of beans, 4,000 tons of amaranth, 2 million cotton cloaks, and huge quantities of cacao beans, war costumes, shields, feather headdresses, and amber, all of which was stored in central warehouses and distributed by tribute officials. Cacao beans served as a kind of universal currency among Mesoamerican groups.[8]

The fragility of the food supply spurred the conquests of the Aztec rulers in an effort to extract ever more tribute. In the early 1450s the Aztecs suffered horrible famines after locusts and floods reduced their crops. Parents bartered their children for corn, and people sold themselves into slavery to survive. The ruler Moctezuma Ilhuicamina (r. 1440–1468) proclaimed war as the central occupation as a way to provide a reliable food supply.

At their beginning in 1325 the Mexicas were a kin-based tribal society. As they built Tenochtitlan they rapidly transformed themselves into a highly stratified urban civilization, with a tiny minority of nobles and priests who were served by commoners (farmers, artisans, fishermen), serfs, slaves, and captives. Nobles were permitted polygamy; others practiced monogamy. Clothing indicated rank and status. Only nobles could wear cotton, traded from the tropical lowlands; everyone else wore fabric woven from the fibers of the maguey (agaves) or related plants. Capes were the major status garment; the length, fabric, and decoration indicated exactly one's family and rank. War leaders wore elaborate feather-covered tunics and headdresses whose colors and styles represented coyotes, jaguars, and death-demons.

Every Aztec belonged to a *calpulli* (big house), a group of families that claimed descent through the male line from a common ancestor. Neighborhoods were based on the calpulli, which could contain thousands of people; each one paid tribute, raised soldiers, had its own temples and schools, and held land in common. The close ties of the family and the calpulli provided the framework for each person's life from birth to death.

The Aztecs called their epoch, beginning in 978 CE, the Fifth Sun; they believed that the gods had destroyed the world four times previously and that their world would be the last, to be destroyed by earthquakes after one of the fifty-two-year cycles. (The Aztecs used the dual calendar developed by the Olmecs, Mayans, and Toltecs.) Only human sacrifice could propitiate the gods, they believed, into postponing the final end; their chief god required a daily diet of human hearts. Their culture was distinguished by the scale of human sacrifice, something that began with the Olmec practice of bloodletting to the gods. But the Aztecs, seeing themselves as carrying the burden of warding off the end of the cosmos, transformed a limited practice into the central element of their ideological system.

In the process of doing this, the Aztec promoted the worship of Huitzilopochtli (literally, hummingbird on the left), the god of war, over the worship of Quetzalcoatl (feathered serpent), the god of agriculture and the arts. Huitzilopochtli had earlier been a minor god among the hundreds worshipped, but the Aztecs, or possibly the Toltecs before them, promoted him to be the Sun God, whose chosen people they believed they were. Every morning at dawn quail were slain in the temples and incense offered to his honor.

From the ages of twelve to fifteen, all Aztec children attended the House of Song, attached to the temple, which provided a curriculum of singing, dancing, and music. In addition, boys were trained in warfare. Noble and commoner boys attended separate schools for military training; the noble boys grew tufts of long hair on the backs of their heads which could only be shorn after they captured their first warrior in battle. Since captured warriors were sacrificed to the Sun God, Huitzilopochtli, Aztec warriors tried to capture prisoners rather than kill them. They used javelins and wooden clubs edged with obsidian and carried wicker shields covered with hide and decorated with feathers. The annual wars began after the harvests were in. War served as a means of increasing tribute of food and of feeding the Sun God with human blood and hearts.

The sacrifice of human victims took place at sunrise or sunset. The priests painted victims with red and white stripes, reddened their mouths, drew black circles around their mouths, and glued white down on their heads. The priests escorted the victims up the steps of the pyramid, thrust them backward over the sacrificial stone, and held down their four limbs while the presiding priest opened their chests with blows of a knife (chalcedony or obsidian). Then he thrust his hand into the chest and pulled out the still-beating heart, held it high, and flung it into a special bowl. The priests tipped the body over to tumble down the steps, possibly to the waiting captor, who may have taken it home.

How many were sacrificed? No reliable figures are available. The Spanish conqueror Cortez estimated fifty at each temple annually, which would have amounted to some 20,000 persons sacrificed per year in Aztec territory. But the Spanish were so horrified by this practice, which seemed to them a grave sin, that they probably exaggerated the number.

The Aztecs believed that warriors who died in battle and women who died in childbirth earned the most glorious life after death. Warriors who died were believed to journey with the sun for four years, then return to Earth as hummingbirds. Women who died in childbirth, after journeying with the sun for four years, returned to Earth as female goddesses. Other people journeyed after death for four years to the place of the dead, the void inhabited by the ancestors. The journey could be easy or terrifying, depending on one's wealth and obedience to the rules of life.

The Aztecs developed a form of writing that used glyphs, or pictures that represented objects themselves, and hieroglyphs (also called pictograms), standing for words or ideas, such as a bundle of reeds to symbolize the fifty-two-year cycle or a scroll with flowers to mean poetry or song. Aztec writing was designed for keeping records and as a prompt for speechmakers; it did not convey human speech itself, just the general sequence of topics and ideas. The glyphs were painted onto paper made by soaking the inner bark of fig trees and beating it to create long strips that were folded into accordionlike books. Scribes wrote on both sides in bright colors of red, yellow, blue, and green. The size of the figures and many details of color and decoration conveyed information in an elaborate code. Some phonetic symbols were used for personal names and places. The Spanish friars (traveling priests) who arrived in the sixteenth century burned hundreds, if not thousands, of these priceless books.

How do we know anything about Aztec life? Two first-person accounts were written by Spanish conquistadors who arrived in November 1519—five letters by Hernando Cortez to his emperor written from 1519 to 1526, and an account by Bernal Diaz del Castillo written almost fifty years afterward. In 1542 a Catholic priest, Bartolomé de Las Casas, wrote an account of the destruction of Aztec culture to protest the cruelty of the Spanish. But the most important written source remains a twelve-volume work by the friar Bernardino de Sahagun, who arrived in Mexico in 1529, learned Nahuatl, and spent years interviewing Aztec elders, who produced hidden documents to help them give talks recalling Aztec history and culture, on which Sahagun took notes. His book was written from 1547 to 1569, but most of it was not published until the nineteenth century, since religious authorities in Europe considered the material too offensive. Many other scholars and archeologists have contributed to what is known of the Aztecs, including excavations after 1978 of the ruins of the Great Temple under Mexico City.[9]

Urban Centers in South America

The natural setting of western South America, with its compact vertical arrangement of landforms, is unique in the world. Over the last 40 million years the Nazca plate at the bottom of the Pacific Ocean has been sliding eastward under the South American plate, raising high mountains close to the coast and creating a deep trench in the ocean floor off the coast. Two parallel mountain ranges dominate the Andes mountains; they rise so rapidly from the coast that the continental divide lies only 60 miles (100 kilometers) east of the shoreline at Lima. Compressed into this narrow vertical band are a coastal desert that goes years without rain, the highest peaks in the western hemisphere, and dense tropical jungles, all of which creates a mosaic of microclimates and distinct ecological belts less than an hour's walk apart. Successful human inhabitants of such an environment had to have intimate knowledge of the many landscapes and their life-forms.[10]

Tropical areas in South America could not produce enough storable surplus food to feed urban populations. Hence, these developed only along the western coastline of what is now Peru and in the nearby Andes mountains. Our knowledge of these South American societies remains inferior to our knowledge of the Aztecs, since these societies had no traditions of writing, not even a glyphic mnemonic system.

What is known of these South American societies is based on records set down by the invading Spaniards and by later native descendants of the

Inca, and by archeological excavations, which began seriously in the 1970s after much looting had already taken place. Two proud accounts of the Inca past appeared early in the seventeenth century written by sons of mixed Spanish-Inca ancestry, one the illegitimate son of a Spanish conquistador and an Inca princess.[11]

By 4550 BCE people in the valleys of the Andes mountains were cultivating potatoes and quinoa. Andean people domesticated two animals that carried loads and provided protein—llamas and alpacas. Llamas could carry about 70 pounds (32 kilograms) each, and one man could manage a group of 10 to 30 llamas. People on the coast, who in the early days seemed to have had little contact with people in the mountains, relied heavily on fish; by 3000 BCE they were producing cotton textiles, and by about 1800 to 1500 BCE, pottery. A primitive form of corn was introduced from Mesoamerica in about 3000 BCE to both the Andes and the coast.

Small towns grew on the coast and in the highlands, each with distinctive, vibrant cultures. By about 400 CE two small cities were thriving—Tiwanaku near Lake Titicaca and Wari (both Quechua spellings). They declined at around 1000 when people apparently left them to return to rural hamlets, for reasons unknown.

Following the decline of Wari and Tiwanaku, the Inca were just one of several ethnic groups in southern Peru hoping to extend their power by warfare and marriage alliances. Sometime around 1400 the Inca began to emerge as the dominant regional group, and within a few decades a society of about 100,000 people asserted its rule over a population of 7 to 12 million—an astonishing feat, possibly paralleled only by that of Alexander the Great. Sometime in the decades of their expansion, the Inca rebuilt their city of Cuzco, nestled in a high mountain valley, to be the sacred center of their empire.

In 1438 the Inca repulsed an attack from neighbors; their defense was led by the king's younger son, who, after his success, renamed himself Pachacuti, meaning Earth-shaker or World Transformer in the Quechua language. Pachacuti repudiated his father and his older brother to begin the Inca empire, followed by his favorite son and successor, Topa Inca. Together, within forty years, they brought into tribute people living 2,500 miles north to south and several hundred miles west to east, from coast to jungle (Fig. 9.2).

An empire must have some system of transporting goods. The Inca constructed a network of roads that rank among the world's great achievements of public works. Built of stone in mountain terrain, the roads stretched for more than 15,000 miles (ca. 25,000 kilometers) and served

9.2   The Inca Empire in 1532

as travel routes for imperial messengers and armies, and for pack trains of llamas.

The state functioned without a tribute of goods; instead its revenues were based on labor service given by its subject peoples, who tilled its fields, cared for its herds, made its cloth, and built its roads, bridges, and cities. There was no private ownership of land; each clan, or *aylla*, held land communally and raised its own food while providing a yearly quota of workers to the state. The state commanded enormous wealth and was expected to supply its workers generously with food, drink (corn beer), and music while giving gifts of cloth to officials, the army, and to people newly incorporated into the empire. The society was sharply class conscious, even though there was no money or private ownership.

Inca farmers domesticated dozens of plants, including cotton, potatoes, and quinoa. They developed the primitive corn introduced from Mesoamerica and used it for brewing beer. For meat the Inca consumed llama and alpaca, both domesticated, and wild vicuna. They grew coca; only the ruling class was permitted to chew the leaves for a small amount of cocaine.

The Inca achieved a distinctive excellence in the weaving of fabric. Cloth held such importance that it was burned as a sacrifice. People of each nation and province had insignias and emblems by which they could be identified. Tunics encoded distinctions of status; headdresses indicated ethnicity and class. Weddings and funerals required special fabrics; diplomatic and administrative protocol depended on textiles. More than 500 hours of labor could go into spinning and weaving a single poncho; the state had a special category of unmarried women, called *mamakuna*, who were devoted full-time to spinning and weaving fabric for state use. For fibers they used wool from llamas, alpacas, and vicunas, and native cotton, cultivated in five natural shades from white to dark brown. They also used fibers from certain plants for added strength and decorated their fabrics with feathers and plaques of gold and other metals.

Inca law restricted luxury goods and precious metals to the emperor and nobles, who controlled mining and metal production. Tin and bronze were most common; since most gold and silver objects were carried off by the Spaniards to be melted, their role in the culture cannot be described.

As the only major agrarian civilization without writing, the Inca developed other techniques to record and transmit information. The best-known one is called the *quipu* or knotted strings. Others include painted sticks, illustrations painted on boards, and designs woven into fabrics.

The quipu came from traditions that went back almost 1,000 years before the Inca. Only about 400 quipu remain, since the Spanish destroyed those they found. A quipu consisted of a primary cord, usually of cotton but sometimes of wool, to which a series of knotted cords were tied. Different kinds of knots indicated different numbers, based on a decimal positional system. The knots were also dyed in hundreds of colors.

The complete meaning of the quipu has not been fully recovered. Each one was accompanied by an oral account memorized by the knot record keeper, who held a professional position under the Inca. The quipu were used to record numerical data from census records, counts of flocks, tax obligations, and warehouse contents. They are thought to have somehow aided in keeping genealogical records and in remembering poetry. Common people used them to keep track of community herds, a practice that continues today.

The Inca beliefs were animistic; they felt that places and natural objects were spirited and sacred. There were few high-status priests and not many ceremonial sites of great scale and elaborateness. Many deities were worshipped, especially three intertwined ones: the Creator, the Sun God, and the God of Thunder. The official religion centered on the worship of *Inti*, the Sun God. The high priest of the sun was usually a close relative of the ruler. The moon (*Mama-Quilla* or Mother Moon) was considered the wife of the sun, gold the sweat of the sun, silver the tears of the moon.

The Inca practiced some human sacrifice, for special solemn occasions such as the ascendancy or death of an emperor, or when an earthquake, epidemic, or eclipse struck. The sacrificial offerings were sometimes prisoners of war but more often were ten-year-old boys and girls chosen for their beauty. Each town had to send one or two pairs of children to Cuzco, where they were sacrificed in the main plaza by strangling or by throat cutting. The rationale for this is not clear, but chroniclers say the intent was to send humanity's best to join the gods and to accompany rulers into death.

Apparently the Inca believed that after death the spirit stayed on, inhabiting the land, requiring drinks of *chichi*, the fermented brew made of corn or other plants. After the death of a ruler, the Inca made a mummy of the body that, with the help of assistants, carried on as if the spirit had never left. Royal mummies were kept in their own houses; they ate, drank, urinated, visited each other, and sat at councils, attended by a special group of their relatives. The bodies were mummified by removing the internal organs, the major muscles, and sometimes the brain. The body cavity could be filled with ash and coal and allowed to dry out. The joints were bound, the spine made rigid with cane supports, and the body filled with feathers, grass, shells, and earth.

One of the archeological splendors of the world is the Inca site at Machu Picchu (meaning old hill), located in Peru on the lower Urubamba River about 45 miles (75 kilometers) northwest of Cuzco, never discovered by the conquistadors. This site came to the world's attention in 1912, after local farmers guided Hiram Bingham there in 1911. Bingham led archeological work at Machu Picchu under the auspices of Yale University and the National Geographic Society and published accounts of it in 1913 and in 1930. Set on high cliffs in the jungle, Machu Picchu consists of complex terraces and elegant architecture laid out around a main plaza. The architecture incorporates immense natural stones. Rather than a religious ceremonial site,

Machu Picchu seems to have served as a royal vacation retreat for the founding emperor, Pachacuti. The empire he founded had less than a hundred years to develop before the arrival of people from across the sea.[12]

The Rest of the Americas

While agrarian civilization began to develop in Mexico and the Andes, the rest of the people of North and South America remained semisedentary hunter-farmers or nomadic hunter-gatherers. The environment in many places limited more intensive agriculture and kept the population at a level that could be maintained by hunting supplemented with limited farming.

Only in a few areas did modest ceremonial centers arise, especially in the rich river bottomlands of North America. These apparently occurred first in what is now Louisiana at about 1000 BCE, then from 500 BCE along the banks of the Missouri River. Called the Hopewell culture (named for the farm on which the first artifacts were found), these people of the Ohio River valley domesticated seed crops and built monumental earthworks. Their towns were hierarchical, led by a chief, with a maximum of several thousand people. They raised corn for beer and traded by river networks to areas now known as Wyoming for obsidian, Lake Superior for copper, North Carolina for mica, and the Rocky Mountains for the teeth of grizzly bears. The Hopewell sites were abandoned in about 400 to 500 CE, attacked by raiders from the north armed with bows that may have come to North America with the Inuits three or four centuries earlier.[13]

The Hopewell culture continued in what is called the Mississippian culture, 700 to 1500. People along the Mississippi River grew corn and squash brought from Mesoamerica. They erected a city called Cahokia near the present East St. Louis, Illinois, where the Missouri River joins the Mississippi, which is now North America's largest mound, about a hundred feet high, containing elite housing and temples. The city reached the height of its population, some 30,000, in about 1200, but was abandoned some fifty years later for no known reason. One chief's burial site from this area contained the bones of more than fifty young women and retainers.[14]

In the desert southwest of what is now the United States, immigrants from farther south brought irrigation agriculture by about 300 BCE. This enabled the population to grow, and settled village life appeared. The Hohokam people of the Salt and Gila river valleys show the strongest Mexican influence, complete with ball courts. Farther north, in the Four Corners area of Arizona, New Mexico, Colorado, and Utah, the Anasazi lived in large villages, growing corn, beans, and squash, from about 450 CE on; after 900 they constructed large residential and ritual centers. Drought probably forced the abandonment of the larger towns in the twelfth century, with people retreating to hard-to-reach caves high above valley floors, suggesting increased warfare provoked by population pressure on limited arable land.

Across the Americas lived innumerable small groups of people, hunting and farming or gathering, creating dazzling cultures and arts of their own, trading by water transport and human portage. The bison hunters of the plains, the Inuits of Alaska, the Tainos in the Caribbean, the people of the Amazon Basin—each integrated themselves into their place and their environment, creating human societies of amazing variety and beauty. The conditions for the development of empires simply did not exist among these semisedentary societies. The food base was not adequate for accumulating vast surpluses; communities, although ruled by a chief, practiced consensus politics; chiefs could not demand tribute payment or labor from their people. Some intergroup alliances developed, like the Iroquois Confederation of Five Tribes, but even these functioned more as nonaggression pacts than as a way to provide policy or military operations.[15]

The food base did, however, sustain in many places healthy, well-nourished people. Across North America people created a "fast food," called pemmican, which provided a complete diet: dried smoked meat (bear or buffalo) pounded into shreds, with animal fat and berries added and pressed into bars. When the Pilgrims arrived, they found themselves markedly shorter than their native hosts.[16]

The Americas in the Context of Afro-Eurasia

The Americas provided an arena in which human societies could develop entirely independently of those in Africa and Eurasia. When we compare the development in both places, we observe remarkable similarities. Tentatively we can conclude that human history evolved along parallel paths in two separate hemispheres, with little or no contact between them.

As people in both hemispheres of the world turned to agriculture, one earlier than the other, they needed first the services of priests, and then those of warriors in order to thrive. The priests helped people learn when to plant and how to save enough seed for the next planting. They did this by observing the heavens to determine times for planting and by rationing consumption through the year by means of feasts, fasts, and sacrifices. The priests controlled a reserve of food to be gifts to the gods, which they could use to support more religious rituals and to relieve famine when it occurred. Farming communities led by priests could fare better during disasters; hence, priests gained power.

Surpluses of food, however, quickly led to organized robbery and the need for professional warriors. Power went to the leaders who could organize fighting forces and protection payments. Hierarchies of elites emerged as food surpluses increased. Military elites entered into coalitions with priestly groups that had closer relationships with ordinary people. This seems to be the common pattern in the development of all human societies.[17]

These patterns can be seen developing in the cultures of the Americas, as centralized empires developed there by the 1400s. American agrarian civilizations developed later than those in Afro-Eurasia, primarily because the plants and animals that could provide surpluses of food simply were not there—no goats, sheep, cows, horses, wheat, barley, oats, or olives. Studying American agrarian civilizations from only 500 or 600 years ago gives us a glimpse into how towns developed into empires in Afro-Eurasia, where this process occurred some 3,000 to 4,000 years earlier and the evidence is scantier.

In Afro-Eurasia a web of interaction and shared learning had been established over thousands of years back to the domestication of pack animals and of horses for rapid travel. These webs were networks in which it was easy to travel and to exchange goods, ideas, best practices, and diseases. These exchanges resulted in more wealth, less cultural diversity, more powerful and hierarchical societies. In the Americas such webs of exchange were beginning but had not reached the density and complexity of those of Afro-Eurasia. As world historians who focus on networks, or webs, J.R. and William H. McNeill write:

> But the world's webs were not equals. The biggest and densest was the Old World Web [Afro-Eurasia]. Its constituent societies included the most formidable on earth, in terms of military and transport technology, in terms of their ability to focus political power at chosen times and places, and in terms of their disease resistance. They may not have been the world's most pleasant to live in—certainly not if one chooses child mortality or social equality as one's indicators—but they were the most formidable.[18]

In the late fifteenth century, the Americas were home to some 40 to 65 million people. (The estimates range from 5 to 100+ million, but they seem to be converging in the middle.) This density was achieved by the domestication of corn and its combination with beans and squash. In addition, the lack of close contact with animals, or with people from Afro-Eurasia, prevented disease on any large scale.

Most of the Americas' people lived in central Mexico, up to 25 million of them. There, at the end of the fifteenth century, the Aztecs, ruling from Tenochtitlan, dominated the scene. Their empire, begun in 1428, lasted a mere three generations. In the Andes lived another 12 to 15 million people, organized by the Inca for less than a hundred years. Both civilizations in the Americas were young and dynamic when the Europeans reached them, building on earlier traditions of urban life that had repeatedly suffered contraction when food supplies had diminished. The rest of the Americas' people—in North America, in South America outside the Andes, in the Caribbean, and in Central America beyond Aztec territory—still lived outside of states run by bureaucrats and governed by laws. In all areas human minds had created a wealth of novel ideas, art, stories, philosophies, religions, and ways of governing.

Leaving the Americas at the end of the fifteenth century, we return in the next chapter to Afro-Eurasia to relate its story from about 1000 to 1490. The following chapter will tell of the fateful encounter of people in the Americas with the Spanish and Portuguese sailors who voyaged over the ocean of time to connect the Earth in its whole circumference.

Unanswered Questions

1. When did people first come to the Americas?

No question has engaged American archeologists more than the time frame for the arrival of the first people in the Americas. Since the shallow sea bottom was exposed from about 75,000 to 12,000 years ago, Siberian hunters could have crossed at any time. But archeologists believe that northeastern Siberia probably remained uninhabited until 30,000 to 40,000 years ago; earlier the challenges of life in the tundra would have been too great. That puts the likely period of crossover sometime between 30,000 and 12,000 years ago.

In 1950 the dating of archeological finds by measuring the decay of radioactive carbon became available, spreading hope that the question of American settlement could be answered. Yet in the years since, the answer is still not clear. There are very few archeological sites older than 10,000 years, and reliable dates have been difficult to establish. The sites with the most persuasive evidence from pre–11,500 years ago are found in Argentina, Chile, and Venezuela, and at Meadowcroft, a rock shelter near Cross Creek in the Ohio River basin of southwest Pennsylvania. The oldest floor in Meadowcroft has been dated at 19,600 years, though some believe this date is not reliable. The debate continues as fresh evidence is uncovered. If people did manage to arrive in the Americas before 9500 BCE, it is clear that conditions were not present for a population expansion to occur until then.[19]

2. How much cannibalism was there among the Aztecs?

Many anthropologists believe that the Aztecs ate human flesh as a religious rite, but nobody agrees on whether it was a symbolic gesture or part of the regular diet. One scholar, Michael Harner of the New School of Social Research in New York City, has suggested that human flesh provided a source of protein for a population living without substantial meat supplies. Anthropologist William Ahrens believes that the Aztecs were not cannibals at all, that the conquistadors' and friars' accounts were biased, written partly to justify their own killings. One of the early friars, Diego Duran, said he saw a thigh cut off and carried away, presumably to be eaten by those who had captured the victim. Only the archeological record, the presence of butchered bones, can provide proof of cannibalism, and so far no one has excavated any Aztec home, where it would have most likely taken place.[20] As the population grew, so did the incidence of war and human sacrifice. The rationale was to prolong a doomed world, which the gods would eventually destroy. Perhaps leaders used human sacrifice as a political strategy, as population control, or as a means to control rebellion. It apparently served as an effort of the entire populace to prolong the Fifth Sun. All these are possible interpretations.

3. What are the Nazca drawings?

One of the mysteries of South America is the designs at Nazca in Peru. These markings on the desert surface are large-scale drawings of animals, straight lines, and geometric shapes. Some are over 2,000 years old; they survive because of the scant rainfall. People made them by removing the thin covering over the soil called desert varnish. This dark varnish consists of manganese and iron oxides deposited over millennia by aerobic microorganisms; people removed it to reveal the lighter-colored soil beneath. The largest designs are several kilometers across; one monkey is 328 feet (100 meters) across, and one bird is over 984 feet (300 meters) long.

No one knows what these markings meant to the people who made them. The geometric ones could indicate the flow of water or be connected to rituals to summon water. The spiders, birds, and plants could be fertility symbols. Other possible explanations include: irrigation schemes, giant astronomical calendars, or landing strips for spaceships.[21]

10

One Afro-Eurasia

(1000–1500 CE)

In the whole world of 1000 CE the agrarian civilizations controlled less than 15 percent of the land now ruled by modern states. Despite the focus on agrarian civilization in most history books, the barbarians (so-called by the urban elites) controlled most of the world's land. The barbarians consisted of foragers, pastoralists, horticulturalists, and small-scale farmers; they lived in the Amazon Basin, North America, western and central Africa, the steppes of central Eurasia, southeast Asia, and Melanesia, constituting a world of economic and cultural heterogeneity.[1]

Our story in this chapter will begin with the Mongols, a nomadic people living in part of the high plains, or steppes, of central Asia, now Mongolia. For the period 1210 to 1350, these people from the margins of urban empires managed to create their own empire which controlled all of Asia from Korea to Hungary, with the exception of India. Without a base of agriculture or urban life, Mongols established the largest continuous land empire in all of human history, lasting about 200 years (Fig. 10.1).

To grasp the land size of the Mongol empire, it is helpful to compare the areas controlled by various states and empires, by using the measurement of 1 square megameter = 1 million km². The Han dynasty of China controlled about 6 square megameters of land, while the Roman Caesars ruled over 4 square megameters. Early Islamic empires of the seventh and eighth centuries controlled 10 square megameters, the Inca about 2 square megameters. The Mongol Khans ruled over 25 square megameters of landmass.[2]

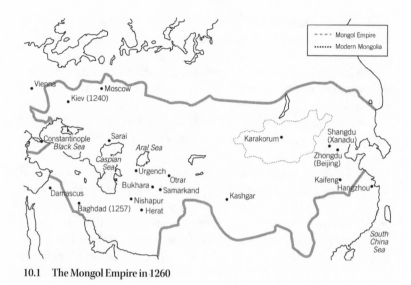

10.1   The Mongol Empire in 1260

The Rise and Spread of the Mongols

From a European perspective the Mongols were usually thought of as hordes of savage men on horseback, riding swiftly into sedentary, urban areas where they murdered and plundered with fearsome cruelty. Europeans called them various names—Tartar, Tatar, Mughal, Moghul, Moal, Mongol. Their reputation continued through the centuries; in the nineteenth century European physicians explained how mothers of the superior white race could give birth to retarded children by claiming that the facial features demonstrated that some ancestor must have been raped by a Mongol warrior (hence the term "Mongolism").[3]

The perspective of Western historians is shifting rapidly, however, as new sources reveal the Mongolian perspective—namely, that the Mongols, and especially their first leader, Genghis Khan,* can be seen as visionaries who incorporated into their huge empire many ideas and values, such as religious tolerance, diplomatic immunity, free trade, and international paper currency, that presaged the modern world.[4]

Primary sources documenting Mongol life and history are scant. No-

* There are at least a dozen systems for transliterating Mongol names; no agreement exists, though there is a trend to use contemporary Mongolian forms. I use the forms most familiar to English readers—i.e., Genghis rather than Chinggis, Karakorum rather than Kharakorum, Kublai rather than Qubilai.

mads do not usually bury large deposits of material that later generations can dig up; there are no archeological sources of Mongol history. Even Genghis Khan's burial place has never been found; his people took extreme care to insure that his body would never be disturbed in the pristine wilderness where he was born, near the present border between Mongolia and Siberia.

Only one primary source exists, an account called *The Secret History of the Mongols*, written by an unknown narrator, probably in 1228 or 1240, just after the death of Genghis Khan or twelve years later. This writer wrote in a Uighur variety of Turkish script, the one chosen by Genghis Khan as the script for the Mongol empire, since previously the Mongols were an oral people without writing. In the fourteenth century this secret history was translated into Chinese script using the phonetic value of the characters to represent Mongolian sounds, and this version is the only one that survived. By the 1980s English versions of this work were available, but not until the early 1990s did Mongolian scholars translate and annotate the text, making it more comprehensible by matching it to the places described.[5]

Life on the steppes of Asia was organized around grazing a variety of animals, primarily sheep and goats for food, and horses, cattle, and camels for transportation. (Horses had been domesticated in the Black Sea steppe lands possibly as early as the fifth century BCE but at least by the third century.) A proper mix of animals had to be attained, and they were consumed as they lost reproductive significance. The size of the human population was strictly determined by the size of the herd. A minimum of fifty to sixty animals was necessary for a family, or a ratio of fifteen to twenty animals per person. A harsh reality prevailed: the loss of animals meant the loss of people.

Mongol subsistence was based on herding the "five snouts," as they called them—cows/yaks, horses, goats, sheep, and camels. From their animals the Mongols had food (meat and milk), clothes (wool, fur, leather), and shelter (felted wool hung on circular frames, called *yurts* in Russian and English, *ger* in Mongolian). The Mongols, however, needed iron for bridles, stirrups, wagons, and weapons, and they eagerly traded for wood, cotton, silk, vegetables, and grains.

The nomads of the steppes had related to sedentary societies for more than a millennium. Their way of life could not be maintained without acquiring from sedentary people a few essentials, like the iron for making stirrups and bridles. The nomads had two choices, to trade or to raid, both of which they employed as circumstances dictated. The rise of the Mongol empire culminated this millennium-old confrontation.

The personal story of Genghis Khan proves to be one of the most amazing in human annals. After a childhood of desperate deprivation, this nonliterate man united all the Mongolian tribes, conquered more than twice as much as any other person in history (whether counted in area, countries, or numbers of people), imposed peace, an alphabet, and religious freedom, and died at almost seventy years surrounded by loving family and loyal soldiers.

Called Temüjin as a boy, Genghis Khan was the son of a minor chieftain who was killed when Temüjin was about nine. His father left two wives (Temüjin's mother, Hoelun, was the second), and seven children. Their tribe deserted them, deciding there was no man available to hunt and provide for them. But, located on the edge of the forest, the women and children managed to survive on their own, living an almost feral existence, hunting small animals, fishing, and gathering berries for subsistence.

When it became apparent to Temüjin that, on coming of age, the slightly older son of his father's first wife would marry Temüjin's mother and take charge of the little group, Temüjin and his younger brother killed their half brother with an arrow each from front and back. This made them outlaws, and they grew up ruthless in a context of kidnappings, murder, and tribal violence, with no formal schooling of any kind.

When Temüjin's own bride, Bortei, was kidnapped from him, he fought to regain her and moved with her out of the forest into the high plains, where he began organizing his own army to defeat other Mongol groups and stop the incessant warfare. He attracted young warriors to his standard, made alliances, and over twenty years systematically defeated all competitors. In order to unite the Mongols, Temüjin neutralized the power of his own relatives, killed the lineages of aristocrats and all rival khans (chieftains), abolished the old tribes and reorganized everyone, and allowed the most powerful shaman in the realm to be killed.

In 1206, when Temüjin was about forty-four years old, he was acclaimed supreme khan of all the Turko-Mongol tribes and assumed the name Genghis Khan. Rather than use any tribal name for his followers, he called them the People of the Felt Walls. As the great khan of the Mongols, Genghis ruled over territory the size of modern western Europe, with about 1 million people and 15 to 20 million domesticated animals.

Mongol people believed in the ancient spirits of Earth, chief among them the Eternal or Blue Sky (*Tengri*). Genghis had an ideology of world conquest granted to him by the "eternal sky" of Tengri. Below the sky lived a host of spirits, mediated to the Mongols by their shamans. As the Mongol empire expanded, its leaders treated other religions with respect, enjoyed comparing the religious ideas they encountered, and practiced religious tolerance within their families.

Genghis Khan organized his army, as was customary in his culture, on a decimal system, down to camps of ten men, who lived as brothers. Every man under sixty was liable for military service. Ten thousand men formed the major fighting unit, while the individual soldier identified with his unit of 1,000, abandoning any tribal loyalty. Commanders were promoted by effectiveness alone, not by birth or position. Every soldier had numerous extra horses; five extra seems not to have been unusual. Soldiers could ride for ten days without building a fire; they ate dried milk paste mixed with water or raw meat prepared by storing it under their saddles to tenderize it. In this way they could cover thousands of miles at great speed. At first soldiers were paid by loot, later by salaries and produce. Families of fallen soldiers also shared in the loot, inspiring loyalty. At his death, Genghis Khan's army consisted of 129,000 men. No general deserted Genghis Khan in six decades of fighting—possibly a unique record!

After unifying the Mongolian tribes, Genghis Khan turned to northeastern China, where a Manchurian group, the Jurcheds, had already taken China's largest city, Kaifeng, a hundred years earlier. Within four years, Genghis conquered the Jurcheds. Then he spent fifteen years subduing the Tanguts in eastern China (closely related to the Tibetans), the Uighurs and Khitans in central Asia to control the Silk Road, and the people in Khwarezm (south of the Caspian and Aral seas) down to central Pakistan, where the heat became too much. Genghis Khan encouraged written and oral stories of the horrors inflicted by Mongol warriors, to motivate other cities to surrender. Unlike other conquerors, Genghis Khan had the enemy aristocrats killed to prevent future wars against him. He destroyed utterly the cities that rebelled after being conquered, down to the foundations of every building, with no creature left alive.[6]

After the death of Genghis Khan in 1227, the assembled Mongols chose his third son, Ogodei, as his successor. The family decided to attack in three directions, west into Europe, southeast in China, and southwest into the Middle East in order to increase the caravan of tribute flowing into the newly built capital, Karakorum. Genghis's grandson, Batu, led the attack in Europe, conquering Kiev and Moscow and reaching the outskirts of Vienna when, in December 1241, Ogodei died, the last survivor of Genghis's four sons. Batu returned to the family enclave to help choose Ogodei's successor from among the grandsons, which took ten years, thus saving Europe from further depredations.

While the Mongol rulers were on the battlefield, their wives often managed the empire. (Among nomadic peoples, women regularly participated in warfare and governance.) When Ogodei proved frequently drunk, administrative power passed to his wife, Toregene, who served as regent for the ten years after his death. Succession became increasingly contested and bitter; one grandson was able to conquer the Muslims and kill their spiritual leader, the caliph in Baghdad, but the limits of empire were reached in 1260 when an army of slaves directed by the sultan of Egypt defeated the Mongols near the Sea of Galilee in what is now Israel.

After 1265 the Mongol empire no longer held together as a whole. When grandson Kublai was selected as great khan in 1265, some family members refused to acknowledge him, and the empire devolved into four sections, each ruled by a different Mongol leader but in contact with one another. Thus far India had not come under conquest from Mongol leaders, but it would in the late fourteenth and fifteenth/sixteenth centuries.

During the days of the Mongol empire, trade and the exchange of ideas flourished, connecting China, the Muslim world, and Europe across the various routes known collectively as the Silk Road. The Mongols established a communication system with a post station every twenty-five to thirty miles, stocked with horses and fodder, which could be used by authorized travelers carrying a medal of gold or silver, inscribed in Mongolian—a precursor to the modern passport. Some cities prospered, others withered under the burden of tribute. A semiworld system within Afro-Eurasia occurred—a single commercial network linking China, southeast Asia, the Indian subcontinent, the Islamic world, central Asia, parts of sub-Saharan Africa, the Mediterranean, and Europe.

Caravans carrying tribute flowed into the Mongolian capital at Karakorum and surrounding areas. Camels and oxcarts carried so much silk that it was used to wrap and pack other items. Silk flowed in many forms, as embroidered robes, rugs, pillows, blankets, and as fabric of more colors than the Mongolian language could identify. Other favorite items that flowed across the steppes included lacquered furniture, porcelain bowls, bronze knives, iron kettles, carved saddles, perfumes, makeup, jewelry, wine, honey, black tea, incense, medicines, and aphrodisiacs.

After Genghis Khan conquered the Jurched in northern China, so much tribute came to him that he agreed to the construction of buildings to store it near the Avarga River near Karakorum. (Usually the Mongols employed only felt tents.) At the installation of Ogodei as great khan in 1229, he threw open the stored treasure in celebration, and everyone had new silk robes; there were so many that the courtiers all wore the same color, a different one each day.

The ease of trade was described by Wang Li (1314–1389), a native of the western region of China, Jiangxi:

> By the time of [Kublai Khan] the land within the Four Seas had become the territory of one family, civilization had spread everywhere, and no more barriers existed. For people in search of fame and wealth in north and south, a journey of a thousand *li** was like a trip next door, while a journey of ten thousand *li* constituted just a neighborly jaunt. . . . Brotherhood among peoples has surely reached a new plane.[7]

The end to this free-flowing commerce, however, lurked within its contents, unknown to all participants. The first outward manifestation of a problem occurred in 1331, when rather suddenly 90 percent of the people of the Hopei province in northern China, in the area where Kublai Khan had built his capital, died of a mysterious disease. Within a year the disease had struck the royal Mongol family at their summer camp northwest of present Beijing, near the Gobi Desert. Within twenty years China reportedly lost a third to a half of its population. The population in China declined from some 124 million in 1200 to as low as 70 million in 1400.

The disease spread out of China at an alarming rate. It reached the Tien Shan mountains in Kyrgyzstan by 1338 and the Black Sea by 1347 via the Silk Road. By 1348 it had entered Genoa by ship and ravaged the cities of Egypt, Europe, and Turkey. By 1350 it had crossed the North Atlantic to Iceland and Greenland. From 1300 until 1400, Europe lost at least 25 percent of its people (Fig. 10.2).

What was this terrible disease? It became known as the Black Death, because after its victims oozed blood beneath their skin, the blood dried and looked black. Lumps the size of golf balls formed in the lymph nodes and burst open; from the Greek word *buboes*, for groin, came the medical term for the disease, bubonic plague. After several days of agonizing pain, the victim usually died. Sometimes the disease attacked the lungs rather than the lymph nodes, and the person drowned in bloody foam, infecting those around by coughing and sneezing.

No one knew what caused this calamity, but people noticed that it seemed to follow the trade routes. In Europe, Christians blamed it on the Jews, who often worked in commerce and who originated, like the disease, in the east. A papal bull from Pope Clement VI was issued in 1348 ordering

---

*A *li* equaled 500 bow lengths; a hundred *li* was maybe thirty miles.

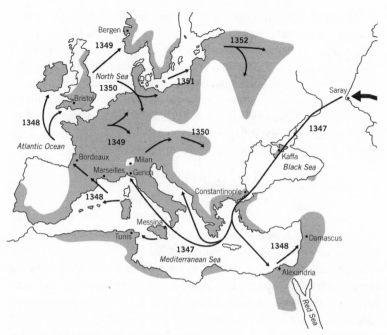

10.2   The Spread of the Black Death
(*Source:* Charles Officer and Jake Page, 1993, *Tales of the Earth: Paroxysms and Perturbations of the Blue Planet,* New York: Oxford University Press, 128.)

Christians to stop burning Jewish people, who fled when possible to Poland, where they were welcomed.

The spread of the Black Death meant the end of the Mongol empire. Trade, its lifeline, dwindled to a trickle. The complex system collapsed without the constant movement of people, goods, and information. Each branch of the Mongol ruling family had to fend for itself without contact with the others. In Russia, the Mongols—there called the Golden Horde, from a Mongolian word meaning court of the khan—broke into smaller hordes that steadily declined in power over four centuries. In Persia, where Mongol rule was called the Il-Khanate, it collapsed in 1335, while in China, the Mongols were defeated and the Ming dynasty was established in 1368. Only in Mongolia and central Asia, called Moghulistan, did Mongol rule continue. By the end of the fourteenth century, Timur the Lame, or Tamerlane, had conquered the Mongol holdings from India to the Mediterranean; his descendants became the Moghuls of India. The last ruling descendant of Genghis Khan, Alim Khan, emir of Bukhara, ruled Uzbekistan until he was deposed in 1920 in the Soviet revolution.

Not until 1894 did scientists understand the true cause of the Black Death and the methods of its transmission. The bacteria that cause the plague, probably originating in the Gobi Desert, live in fleas, which live on rodents. The disease likely traveled on rats in shipments of food. The bacteria then found fecund environments in densely inhabited cities and ships, where rat populations had lived so long in close proximity with humans that no one suspected them as sources of the disease. Today groups of rodents around the world still host fleas carrying the plague bacillus, but antibiotics prevent massive outbreaks of the disease.[8]

Mongols, then Mings, in China

As we have seen, Genghis Khan conquered parts of northern China as his first conquests after he had unified the Mongols by 1206. Successor Mongols conquered more of China, until in 1260 grandson Kublai succeeded in unifying China by toppling the Song dynasty in southern China. Kublai Khan, who ruled until 1294, established the Yuan dynasty, which ruled until 1368.

Most Chinese, of course, hated the Mongols for their uncivilized behavior. (They drank blood, ate raw meat, lived in tents, and wore animal skins.) But Kublai managed to adapt himself to Chinese ways sufficiently to rule effectively. He built a new capital—called Khanbalik (the city of the khan) in Mongolian, Dadu in Chinese—on the site that would later become Beijing. He had his Muslim architect lay out the area that would become the Forbidden City, where he could live behind walls as a Mongol in a miniature steppe—sleeping in yurts, hunting on horseback. Kublai Khan created public schools to provide universal education 500 years before Western governments undertook this responsibility. He established a government printing office that used printing with moveable type, hand-carved on wooden blocks, on a large scale. He encouraged ceremonial dramas that went on for weeks at a time. He abolished Confucian examinations and ranked merchants second only to government officials, with Confucian scholars below prostitutes but above beggars. He asked a Tibetan Buddhist lama, Phagspa, to create an alphabet in which all the languages of the world could be written. (He did, one with forty-one letters.) Under Kublai Khan in the thirteenth and fourteenth centuries, many Chinese technologies, superior to those in other places, were exported through trade and travel: painting, printing, compass navigation, gunpowder weapons, high-temperature furnaces, and perhaps shipbuilding.

Under the Mongols, however, Chinese farmers suffered tax burdens too great to be borne. By the 1360s local rebellions by farmers, combined with feuds among the Mongols, enabled the Chinese to retake their country and establish the Ming dynasty while at the same time preserving the primary Mongol legacy: the unification of China into a country about five times as large as the area in which people spoke the Chinese language.

The first ruler of the new Ming dynasty, Zhu Yuanzhang, moved the capital to Nanjing, away from Mongol territory, and showed distaste for everything Mongolian or foreign. Muslim, Christian, and Jewish merchants were expelled; Mongolian names and clothes were forbidden; Buddhism was rejected and paper money abolished. The Ming government reactivated the Confucian exam system and employed educated men, depriving the commercial world of some of its most ambitious personnel. Only the Mongolian language was kept as the language of diplomacy.

The successor Ming emperor, Zhu Di (r. 1402–1424) moved the capital back to Beijing and rebuilt the Forbidden City in Chinese style. The population began to increase as agricultural production peaked in the mid-1400s. The Grand Canal was deepened to transport rice to Beijing, while a large army protected the northern boundaries against Mongol horsemen.

Between 1405 and 1433, the imperial court of China financed seven major naval expeditions into the Indian Ocean. Six were under the command of the eunuch Zheng He. These expeditions visited Chinese merchants abroad and affirmed the emperor's prestige, but they cost the government dearly. On Zheng He's largest expedition he commanded more than sixty ships and 25,000 to 40,000 men, compared to Columbus's largest expedition of seventeen ships and 1,500 men. Zheng He's largest ships, with nine masts and 500 men, were at least five times the length of the largest that Columbus later commanded. The Chinese expeditions averaged two years each and in total visited at least thirty countries, vividly demonstrating China's superior navigational skills.[9]

Yet the Chinese did not use their naval superiority to continue explorations around the cape of Africa or across the Pacific to the unknown continents. The Ming government decided to focus its resources on internal development and on safeguarding the steppe frontier. It withdrew from its southern expansion into Annam (present Vietnam), allowed its fleet to decay, and prohibited private overseas trade. The ruling elites of Ming society preferred stability to aggression; they managed a successful recovery from the losses that China had suffered under the Mongols. Some scholars argue that the collapse of the trading system protected by the Mongols caused such economic difficulties in China that it had no choice but to

withdraw from the sea and rebuild its agrarian base and internal production.[10] From 1400 to 1700 the Chinese population more than doubled, as did
that of India.

Mongols and Afterwards in the Islamic World

Some historians maintain that Islam, not China, served as the world's most
creative and dynamic civilization from 1000 to 1500 CE, carrying innovations from one society to another, and that an impartial observer in 1500
might well have predicted that Islam would soon become the world's dominant faith.[11]

These assessments stem from three fundamental facts. First, the
world of Islam almost doubled in size between 1000 and 1500; principally
in India (Timur and other Mongol descendants had converted to Islam),
eastern Europe, Africa, and southeast Asia. Second, the elite urban culture of the Muslim heartland (Iraq, Iran, and Azerbaijan), based on Persian-
ized Turkic and Mongol court cultures, flourished for a brilliant period.
Third, Islam served as the hub of the commercial network of Eurasia,
connecting China and India to Africa, the Mediterranean, and Europe
(Fig. 10.3).

The blaze of Islamic culture produced beautiful public buildings, the
Taj Mahal, exquisite illustrated manuscripts, and poets like Omar
Khayyam (d. 1131), Rumi (d. 1273), and Hafez (d. 1389). It also produced an
observatory at Maragha, near the capital Tabriz (now in northwest Iran)
and a mathematician, Nasir al-Din Tusi, who used the observatory to pro-
pose the idea of small circles rotating within large circles that led Coperni-
cus to his insight that the planets circled the sun. Nasir al-Din Tusi also laid
the foundations for complex algebra and trigonometry. (Muslims knew of
the Indian numerical system, which included the use of zero, by the seventh century CE. They used it across their lands, including Spain, where a
French monk who became Pope Sylvester II learned it between 967 and 970.
As pope, he helped popularize it in Europe.)

The agricultural exchange going on in the Muslim world was probably
the most dramatic before the meeting of the eastern and western hemispheres. Arabs brought back from India hard wheat, rice, sugarcane, bananas, sour oranges, lemon, limes, mangoes, watermelon, coconut palm,
spinach, artichoke, eggplant, and cotton. All these spread to Spain except
mangoes and coconut palms.[12]

Eventually, farming retreated in the Muslim world. After 1037 the

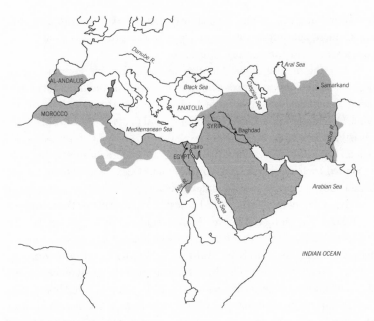

10.3   Islamic Heartlands, c. 1000

Seljuks, who were Turkic pastoralists, prospered enough that they could ride from the steppes into the Muslim areas in Iran and eastern Turkey. Within thirty-five years the Seljuks broke through the Byzantine frontier to occupy most of Anatolia (modern Turkey). This retreat of farming may have been due to the warmer, drier summers that Europe is known to have experienced from 950 to 1250. This weather increased European agricultural production but may have proved too hot and dry in the Islamic world.[13]

The Mongols sacked the Islamic capital, Baghdad, in 1258, seeming to threaten Islamic civilization. But, since the Mongols had no literate culture to bring, the khans assimilated themselves to Persian court culture and converted to Islam in 1295. Ruling until 1353, their regime proved conducive to Islamic growth and expansion.

The khan who converted, Ghazan (r. 1271–1304), was persuaded to do so by the first world historian, Rashid al-Din, a Persian Jew who had converted to Shiite Islam. As Ghazan's prime minister, al-Din also traveled widely and kept in touch with Mongol officials in central Asia and China, advocating monetary reforms that evolved simultaneously in Iran, Russia, and China. As mentioned, al-Din wrote the first history of the world, which included the earliest known general history of Europe, based on information from European monks. He included pictures adapted from European and Chinese paintings, thus introducing Chinese principles of watercolor composition and portraiture to Muslims.

After the demise of the Mongols in 1353, unstable Persian Islamic regimes formed around successful military captains, the most successful being Tamerlane, who ruled from 1369 to 1405. At the same time the Islamic Ottoman Empire emerged from pastoral people on the frontier with Christianity, conquering the northwest part of Anatolia and most of the Balkan peninsula. When land routes were not secure, Islamic merchants expanded their maritime trade in the Indian Ocean and penetrated the Malay-speaking peninsula of southeast Asia.

The extent of the Islamic world is best brought to life by the records of the travels of a Moroccan lawyer, Muhammad ibn Abdullah Ibn Battuta (1304–1368). Ibn Battuta grew up in Tangier, Morocco, the son of a family of legal scholars. After studying law, he set out in 1325, at age twenty-one, to practice law abroad and to visit Mecca and as many other significant cities as possible. He reached as far as southern China and returned home after twenty-four years, then made a two-year trip to Mali before returning to Morocco to write his memoir, the *Rihla,* from memory without notes to help him. In total he may have traveled 73,000 miles (180,000 kilometers), visiting some forty-four modern countries. He worked as a judge in several different places, since all the Islamic countries shared the same sacred law—the Shar'ia. Everywhere he went Ibn Battuta could converse in Arabic with merchants, princes, and scholars on matters of jurisprudence, mysticism, and current events in the "Dar Islam" (the Abode of Islam), confirming the existence of the *umma,* or community of believers.[14]

Along the eastern coast of Africa trade with the Islamic world expanded from 1250 on, giving rise to between thirty and forty city-states—for example, Mogadishu and Kilwa, where people shared a common language, Swahili. Gold exports from eastern Africa expanded during the fourteenth and fifteenth centuries; Kilwa was exporting a ton of it annually by the late fifteenth century. Much of the gold came from or passed through the city called the Great Zimbabwe at its peak at about 1400.

Islamic trade along the Niger River to the Atlantic coast of Africa grew to such an extent that local rulers began to adapt the Islamic faith as a cultural bridge to the larger world. African rulers grew rich exporting slaves, gold, and salt. The kingdom of Mali developed along the Niger River, reaching its height at about 1330; at this time its rulers controlled two-thirds of the world's production of gold. Paper and papermaking were introduced; Timbuktu became the center of learning. Rival local rulers used trading in slaves as a source of state revenue, which limited population growth and agricultural production. As trade declined in the fifteenth century, traders departed and some Africans returned to their traditional animism.

The rising prosperity of the Muslim elites was accompanied by a growth in slavery. Military campaigns in India reduced thousands of Hindus to slavery. In sub-Saharan Africa, local elites enslaved other Africans for sale and, as Muslim customs gained influence, for their own use. By modern estimates, sub-Saharan and Red Sea traders sold about 2.5 million enslaved Africans to Muslim buyers in northern Africa and the rest of the Muslim world between 1200 and 1500, though no figures are reliable. African slaves reached China by at least the seventh century, and by the twelfth century some wealthy people of Canton had black slaves. Some wealthy Muslim men aspired to having a concubine from every part of the known world. One Indian noble reportedly kept 2,000 harem slaves, including women from Turkey and China.[15]

Islam's central location on the Eurasian trading routes connecting China and Europe enabled its ideas and practices to be exchanged with others far and wide. Unlike the Chinese, who rejected Mongol ideas after returning to power, Islamic rulers managed to absorb and adapt their Mongol heritage.

Europe from 1000 to 1500

Europe in 1000 was a rural backwoods, thinly populated, with nine out of ten people living in the countryside. Europeans were called "Franks" by their Muslim and Byzantine neighbors, but called themselves "Latins," for their allegiance to Roman Catholicism and the Latin language used in its rituals. In contrast, a Greek-speaking Roman emperor supported by Orthodox Christianity ruled the Byzantine Empire, which stretched south from the borders of Serbia and Bulgaria across Greece, western Turkey, and southern Italy.

In perhaps the dominant social pattern of western Europe, households of warrior noblemen lived off the labor of fifteen to thirty peasant families, who gave more than half their labor in grain and service to their noble knights in exchange for the use of land and for protection. The peasant farmers worked in plow teams using moldboard plows, which had a moldboard to lift and turn the soil cut by the blade. Mild winters and year-round rainfall meant that agricultural production could increase, with men working three crops a year continuously except for twelve days at Christmas.

One peasant family could manage thirty to forty acres, which, as an example, might produce an average of 10,200 pounds of grain each year. Of this, 3,400 pounds had to be saved for seed, 2,800 pounds were fed to four horses, 2,700 pounds went to the lord, leaving 1,300 pounds for the farmer and his family (amounting to only roughly 1,600 calories each person per day). Hence, the family also had to raise fruit, vegetables, livestock, chickens, and rabbits.[16]

As production rose with better plowing, peasants were able to keep something to exchange for goods. This led to population increasing for three centuries, more than doubling from 1100 to 1445. During this period Europe's once-dense forests were turned to cultivated land, except where powerful owners established hunting reserves.

European politics during the eleventh and twelfth centuries were dominated by the Crusades, a series of Christian military campaigns against Muslims in the eastern Mediterranean. Popes wanted to reform the church and protect its power, upper-class men needed sanction for their armed violence, Islamic forces were advancing on Constantinople, and merchants wanted to increase trade. Incredibly complex alliances arose and fell; the fourth Crusade in 1204, diverted from Palestine, instead sacked Constantinople, the capital of the empire they were to protect, and established Latin principalities on Byzantine territory. A new level of anti-Semitism arose in Europe, and mountains of looted jewels, art, and religious artifacts flowed into European cities.[17]

In Europe the rulers and clerics did not, as in other parts of the Eurasian continent, manage to maintain control over bankers and merchants. Urban self-government prevailed, with plenty of competition, rivalry, and often violence, rather than the peace that imperial states could impose. Neither the German emperor (1250) nor the Roman papacy (1303) was able to consolidate its claim to universal authority; Europe did not become unified but remained ravaged by warfare. The Hundred Years War saw vassals of the French king—namely the king of England and various noblemen—struggle against him from 1337 to 1453. The English captured Joan of Arc, but in the end the French king, Charles VII, won, with both monarchs relinquishing some power to the more representative institutions of Parliament and the Estates General.

During the thirteenth century people in Europe began to get first glimpses of Mongol and Chinese culture. When a Hungarian monk traveled east in 1237 and encountered the Mongol invasion, led by Genghis Khan's grandson, Batu, he had no explanation for who these people could be: One of the ten lost tribes of Israel? Or a plot by the Holy Roman emperor Frederick to gain the homage of the king of Hungary? When the first European envoy to the Mongol court arrived in Karakorum in July 1246, he had ridden for nearly three and a half months, averaging twenty-five miles a day on horseback, to cover 3,000 miles. News could hardly spread frequently or rapidly.[18]

Europe's first hemispheric traveler, Marco Polo, came from a Venetian trading family; he journeyed to China with his father and uncle from 1271 to 1295, a trip possible because of the Mongol policy of permitting merchants of all origins and religions to travel and conduct business. (Polo visited China during the rule of Kublai Khan, whereas Ibn Battuta was there later, ca. 1345–1346.) When Polo's account of his trip appeared in Italy in 1300, many learned men of his time considered it fiction, in part because Polo dictated it to an author well known for his imaginary adventures.

The costs of the Mongol trade system to Europeans would come when the Black Death swept through Europe. Meanwhile, Europe probably gained more than any other area of the world from the Mongol commercial system. By trading with the Chinese, Europeans acquired the tools they would use to gain world dominance in the period after 1500—printing, firearms, and navigational devices. Parchment was replaced by paper, known but rarely used in Europe prior to the Mongol era. By shopping and trading Europeans gained improvements for blast furnaces, new carpentry tools, cranes, and new foods—carrots, turnips, parsnips, and buckwheat. The increase of trade led to the minting of gold coins in 1252 and, by the mid-fourteenth century, to the invention in Italy of double-entry bookkeeping; this made the exact calculation of profit and loss simple for the first time.

From 1315 to 1322 Europe experienced a period of cooler, wetter weather, which resulted in poor crops and widespread famine. The Black Death followed a generation later, from 1347 to 1351, entering at Genoa. (Fig. 10.2). Europe's population in 1400 stood about equal to that of 1200; it did not rise above the plague levels until after 1500. After gunpowder made knights in armor obsolete and the Black Death disorganized everything, serfdom disappeared; the peasants ran away or bought their freedom. Forests recovered somewhat, and the science of forest management emerged in France and Germany in the fourteenth century.[19]

By the mid–fourteenth century Europeans had taken what they learned from the Chinese about gunpowder to develop field artillery. By the mid–fifteenth century Europe's gun manufacturers, funded by private entrepreneurs rather than governments, were using local mining operations to outstrip the world's other armament makers. By 1480 mobile siege guns could breech any castle and, when mounted on ships, could attack other ships and shore fortifications.

During the period 1000 to 1500, learning expanded dramatically in Europe from the days when people lived with just the Bible and reminders of Roman achievements. In the eleventh century Latin Christians took Toledo, Spain, and Sicily back from the Muslims, and they regained southern Italy from the Byzantines. In doing so they acquired the manuscripts of Greek and Arabic monks. In the twelfth century papermaking arrived in Morocco and Spain, having spread from Baghdad to Egypt by 900.

After 1200, new colleges arose in Europe that may have been patterned after the *madrasas*, the endowed places of study spreading in the Muslim world that provided subsidized housing for students and paid the teachers' salaries. Teachers for the new colleges often came from two new religious orders, the Dominicans and the Franciscans. Europeans, moreover, added to the idea of college, by forming universities, which were defined as degree-granting corporations specializing in research and advanced teaching—a significant invention.

Before 1300, Europeans had created twenty universities; they added sixty new ones between 1300 and 1500. In all of them, Latin was used as the language of instruction. Sometimes students banded together to start a university; more often guilds of professors did. The university in Bologna specialized in legal training; those in Montpellier and Salerno focused on medicine; those in Paris and Oxford excelled in theology. Abelard (1079–1142) and Thomas Aquinas (1225–1274), famous professors in Paris, used logical reasoning to find answers to religious and philosophical questions.

After 1450 three improvements revolutionized printing, and hence learning: moveable metal type of individual letters, new ink suitable for paper, and a modified wooden screw press that pressed inked type onto paper. Johannes Gutenberg printed his first Bible in 1454, a book whose beauty testified to his years of experimentation. By 1500 printing presses in Europe were printing annually 10 to 20 million volumes, both ancient texts and contemporary political and religious tracts, in more than a dozen languages.[20]

In Europe private persons could buy guns (power) and books (knowledge). Governments were not able to control change or the continuing commercialization in their societies. This set Latin Europe apart from other Eurasian societies, where more controlling governments were more nearly able to enforce traditional patterns and conduct. In Japan, for example, the manufacturing of guns was restricted and, after 1637, the ruling samurai allowed guns to disappear as not worthy of a gentleman.

The Margins of the Eurasian Core

Europe shared with the countries on the Pacific coast of Asia its marginality to the central Eurasian trading core. Some parallel developments are evident, especially the development of seagoing vessels and navigational skills, preparing the way for full globalization. Malay sailors and merchants extended trade into the more remote islands of the Pacific, among them the Moluccas, Borneo, and Mindanao in the Philippines. Japan managed to fend off Chinese expeditions and to develop its own distinctive culture. Korea and Annam (Vietnam) remained more in the shadow of China but not directly under the imperial bureaucracy. Hence, some rivalry and innovation became possible, as in western Europe.

In Africa one of the world's great migrations took place, as the black-skinned Bantu people began moving out of eastern Nigeria in West Africa in the third or fourth century CE. What set off this migration is unknown—possibly an increase of population caused by an influx of people fleeing the desiccation of the Sahara. The Bantus forged iron, and their iron weapons gave them an advantage over the neighboring hunters and gatherers. Whether the migration was peaceful or violent is not known. The Bantus moved first into central Sudan, then by the thirteenth century into the forests of central and West Africa, to the east coast, and south of the Zambezi River into southern Africa, a gradual migration of 1,000 years.

The grasslands of Africa, from Senegal to Lake Chad, formed the southern edges of the core trading area of Afro-Eurasia, with the Mali and its successor empire, the Songhai, enriched by their trade with the Arab world. But farther south, in central and southern Africa, people remained on the margins of trade. The lack of pack animals and navigable rivers, plus the presence of lethal diseases and recurrent droughts and famines, resulted in few cities and little long-distance trade. Here Africans continued their traditional life centered on ancestors, animism, and local chiefdoms. (The city known as the Great Zimbabwe proved a fleeting exception.)

The Arctic north—Siberia northward, Alaska, northern Canada—also remained marginal to trading; here hunters, fishermen, and gatherers pursued their traditional ways. In Russia the climate proved slightly warmer, with navigable rivers, and there the state of Muscovy rose in the fifteenth century, centered in Novgorod and specializing in fur trading.

By 1500 the world's global population had reached 400 to 500 million. China and India each had about 20 percent of the world's population, as did all of Africa. Europe, excluding Russia, had about 15 percent, while the Americas had something less than 10 percent.[21]

During the five centuries from 1000 to 1500, the people of the core areas of Afro-Eurasia traded, interacted, invented technology, exchanged ideas, mobilized efforts, and managed to increase the power and wealth of their areas to unprecedented heights. The empire created by Genghis Khan imposed the peace that facilitated this exchange, until the empire broke up from internal disputes and the devastation of the Black Death. With trading routes across central Asia no longer secure, the people of southern Europe sought new ways to trade with the people of China. When they succeeded, a whole new chapter in world history began.

Or so Western historians usually tell the story. But it could be said, perhaps more accurately, that the new chapter of the modern world began in the period 1000 to 1500, when the food supplies of all Afro-Eurasia were exchanged, as well as the technology, the inventions, and the socioreligious ideas. Chinese, Indian, and Islamic cultures played leading roles in this exchange, while Europe played catch-up from its location as a backwater. No wonder Europeans were excited by what Columbus did.

Unanswered Questions

1. What is the meaning of "feudalism"? Does this term have any use as a tool for the analysis of world history?

World historians are currently debating this question. If feudalism is defined as a change in the kinds of horses used, in the relationship of warrior-elites to their states and societies, and in the mode of peasant production, some scholars find similar changes throughout Eurasia while others do not. Even using "feudalism" to describe the process restricted to Europe from 900 to 1200 is more complex than an idealized version of this term implies. The warrior aristocracy increased its power and prestige, managing estates on which peasants increased productivity to the point that paid service began to replace unpaid service in exchange for land. At the same time, areas where people still hunted and gathered or lived in tribal peasant groups (stateless areas) were assimilated into the core agrarian lands under the limited state authority. Since the feudal process in Europe involved many levels of change occurring simultaneously, the term "feudalism" must be used with careful distinctions and definitions.[22]

2. Can one-half of 1 percent of the world's male population be descended from Genghis Khan?

Yes, say geneticists, who are now able to track the Y chromosomes that seem characteristic of Genghis Khan and his male descendants. Geneticists in China, Pakistan, Uzbekistan, and Mongolia, led by Dr. Chris Tyler-Smith of Oxford University, collected blood for ten years from populations living in and around the former Mongol empire. They found a signature cluster of Y chromosomes common within the empire and absent outside it, except for the Hazara people of Pakistan and Afghanistan, who are former Mongol soldiers claiming descent from Genghis Khan. The geneticists believe that the signature cluster comes from Genghis Khan and his ancestors, but they cannot prove it because his body has never been found. They also think the signature cluster became common in part because of rapes occurring during conquests, but more often because the khans had access to large numbers of women in the territories they ruled. In the area that used to belong to the Mongol empire, up to 8 percent of living men were found to carry the signature chromosomes; in present-day Mongolia about 20 percent of the men carry the chromosomes.[23]

3. Why did Ming rulers withdraw China from a worldwide navigational role after establishing a commanding lead?

Earlier Western historians tended to consider China's withdrawal from worldwide trading as a major miscalculation, a mistake that cost it the world dominance achieved by Europe after it connected the globe. Current historians tend to argue that withdrawal was a sensible decision by the Chinese government—that with a huge, land-based empire, there was no reason for trying to rule distant colonies. Some see the sailing expeditions as a pet project of Zhu Di; after he died, subsequent emperors had no interest funding huge voyages. Still, this remains one of the big "What ifs?" of world history: What if Chinese ships had colonized the New World?

Recently a former submarine commander of the British Royal Navy, Gavin Menzies, claimed that Chinese ships under Zheng He did explore the Bahamas and the Falkland Islands and did establish colonies in Australia and New Zealand, Puerto Rico, Mexico, California, and British Columbia between 1421 and 1423. Menzies's book, *1421: The Year China Discovered America* became a bestseller in the United States in 2003, but world historians have denied its accuracy, calling its assertions preposterous and its evidence spurious.[24]

11

Connecting the Globe

(1450–1800 CE)

During the sixteenth century people sailing the seas connected the two hemispheres of the world. Seagoing travel had been expanding steadily, and some attempts had been made to cross the unknown ocean, which turned out to be two oceans divided by a continent. The Vikings had gotten to Newfoundland in 1001; Polynesians may have reached the Americas earlier; Mansa Muhammed of Mali may have led an Atlantic seagoing expedition; Basque fishermen caught cod off the coast of Newfoundland sometime in the fifteenth century. But as events transpired, it was sailors from Portugal and Spain who made the voyages that connected Eurasia and Africa to the Americas in a permanent way. The Portuguese and Spanish had the best location for reaching the New World, and they had the resources to establish themselves once they landed.[1]

World historians, from Karl Marx to David Christian, agree that the coming together of the two world hemispheres stands as one of the most significant moments in the history of humanity. It proved brutal and destructive for three of the world's areas—the Americas, Australia, and the Pacific Islands. Eurasian societies prevailed and, as Karl Marx wrote, "World trade and the world market date from the sixteenth century and from then on the modern history of Capital starts to unfold."[2]

The Crucible for Columbus

By 1500, less than 20 percent of the world's landmass was marked off into states run by bureaucrats and governed by laws. The rest was organized into chiefdoms and tribes, most of which had settled down into some kind of agricultural existence. Hunters and gatherers probably constituted only 1 percent of the world's population of 461 million.[3]

Whether in complex states or under local chiefdoms, people wanting to trade organized trading routes and circuits. One such circuit operated among the Pacific Islands, where possibly a few million people with extraordinary maritime skills (without magnetic compasses, using only their observations of the waves, stars, currents, and evidence of land) traded among the islands: Yap in the Carolines, Guam, Palau, Fiji, Samoa, and Tonga.

Another trading circuit existed in the Americas, involving 40 to 60 million people. It linked the Great Lakes to the Andes mountains using overland relay trade and canoes on rivers and among the Caribbean Islands, and connecting with the Aztec system in central Mexico and the Inca empire with its roads in the Andes.

The third trade circuit involved three-quarters of humanity, about 260 to 300 million people, and extended across Eurasia and the northern parts of Africa. It consisted of two main routes: the caravan routes called the Silk Road across central Asia, and the sea routes from Korea, Japan, and China, around the Malay peninsula and the Molucca or Spice Islands, into the Indian Ocean to ports in the Persian Gulf and the Red Sea, extending into Europe by the Rhine, Elbe, Danube, and Po rivers, and into Africa by camel caravan.[4]

Within the third trade circuit Europeans had taken aggressive measures against Muslims, both in the various Crusades and in the efforts of Portuguese and Spanish Christians starting in 1031 to regain their land from Muslims. (Spain still consisted of Aragon, Castile, Navarre, and Granada.) By 1250 Christians had taken all of Spain except Granada, a narrow strip across most of the southern coastline.

During the Crusades Europeans first tasted sugar in Syria; since the European climate (except Sicily) proved too cool for growing it, they had a motive for looking for places capable of its cultivation. In addition, they wanted to tap into a larger share of the spice trade from the east.[5]

European rulers also wanted to find new sources of gold to expand their economies and to underwrite their activities. Most of their gold came from West Africa, from present-day Ghana, Benin, Togo, and Guinea, then called the Gold Coast by Europeans. This gold had to travel by camel caravan through the Sahara to Fez in Morocco or to Tunis or Tripoli, and Muslim merchants monopolized this trade.

Atlantic Europeans had the sea. They took advantage of it by designing oceangoing ships capable not only of sailing the high seas anywhere but also of carrying heavy cannon. The Portuguese developed the caravel, a small, two-masted ship one-fifth the size of the largest European and Chinese ships, with planks nailed to a skeletal rib rather than overlapping each other, as was common in northern Europe. With the caravel the Portuguese gained maneuverability by using triangular sails, called lateen sails (also used in the Indian Ocean), instead of the square European ones; these sails enabled the caravels to sail against the wind. Later, caravels developed into three-masted ships rigged with both lateen and square sails.

Building seaworthy ships solved only half the problem; knowledge about navigation also had to be built up, combining the traditions of Arab astronomy and mathematics with the practical experience of sailors. The key instruments were the magnetic compass, first developed in China, and the astrolabe, invented by the Arabs or Greeks, that measured the position of the sun or the stars to show ship pilots their latitude. In Portugal the king's third son, Prince Henry the Navigator (1394–1460) set up a research institute for studying navigation and collecting maps. He did this after leading Portuguese soldiers in an attack in 1415 on Morocco that failed to defeat the Muslims guarding the inland gold, motivating Henry to figure out how to sail down the coast of Africa. While the Portuguese worked on this, Ottoman Muslim forces took control of Christian Byzantium in 1453, henceforth called Istanbul, rendering the overland trade more difficult and upping the ante for finding a sea route to China.

By the time Henry the Navigator died in 1460, Portuguese sea captains, supported by state sponsorship, had reached islands off the coast of Africa (the Madeiras, the Azores, the Cape Verde Islands) and had sailed down the coast as far as Sierra Leone. In 1487 a Portuguese expedition sailed westward and never returned. By 1488 Bartholomeu Dias rounded the tip of Africa, and in 1497–98 Vasco da Gama led a Portuguese expedition around Africa to India. In 1500 Pedro Cabral led ships that hit the coast of South America after they swung west to pick up winds they hoped would sweep them around the tip of Africa; this gave the Portuguese claim to Brazil. By 1510 the Portuguese had won a naval battle using cannon onboard ship and had begun establishing themselves in the Indian Ocean.

Meanwhile, the Spanish ruling houses were occupied with driving the Muslims out of Granada. Ferdinand of Aragon and Isabella of Castile, who had married in 1469, united their kingdoms in the 1480s, renewing their determination to defeat the Moors (Muslims). They revitalized the institution of the Inquisition, set up by the priest Dominic to root out heresy in Aragon.

Under the Muslims and under Christian kings before the plague broke out in 1348, Christians, Jews, and Muslims had lived in relative harmony in

Spanish kingdoms. Jews and Muslims had come to own some of the most productive farms and businesses. With the first executions under the Inquisition beginning in 1481, wealth confiscated from Jews funded the war against the Moors, who surrendered early in 1492. Ferdinand and Isabella then established Christian rule with no toleration for other religions. All Jews were ordered to convert or face immediate expulsion. Ten years later Muslims were given the same choice.[6]

In the fateful year of 1492 Ferdinand and Isabella, after their success in driving out the Muslims, finally agreed to sponsor Christopher Columbus's expedition to sail westward to seek China. Columbus had been petitioning for four years without success as the Spanish sovereigns battled their enemy. A climate of Christian militancy victorious after centuries of fighting holy battles against Muslims formed the crucible out of which Columbus and his men sailed on August 3, 1492, from Palos, the seaport near Seville. Columbus was not permitted to get under way until the last Jews had been expelled on August 2, 1492; the Jews sailed away to Portugal, northern Italy, or Holland, or to tolerant Muslim countries in northern Africa.

The roots of Europe's racist thinking had taken hold in Iberia during the intensification of the conflict with the Moors. Most historians find that no concept equivalent to "race" existed in the thinking of Greeks, Romans, or early Christians. There was a hatred of Christians for Jews over responsibility for the death of Jesus, and this intensified in popular opinion during the years of the Crusades. There seems to have been no hatred of blacks in medieval Europe except in Iberia, where Christians learned from Muslims to associate blackness with slavery. (Muslims had both black and white slaves, but generally gave black ones the more menial jobs.) Whites as slaves in Europe declined as European polytheists were converted to Christianity.

Soon after the Iberians drove out the Jews and the Muslims, they enacted purity of blood laws (*limpieza de sangre*) to keep those mixed with Jews or Moors out of some offices, and out of becoming conquistadors or missionaries, who had to be of pure Christian ancestry. Out of these attempts to maintain religious purity grew later European racist thinking in biological terms.[7]

First Encounters

Columbus sailed for Cathay (China) carrying a printed copy of Marco Polo's travels with him. He also took along a man who spoke Arabic to help him

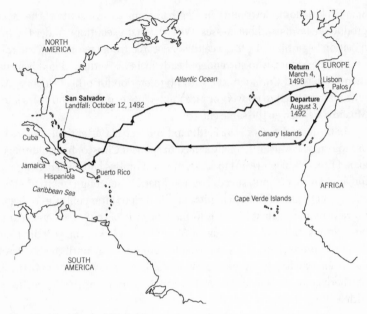

11.1   The First Voyage of Columbus

communicate with the Mongol khans, whom he believed still ruled China though the Mongols had been defeated in 1268.

Columbus made four trips to the Caribbean. On the first he arrived in Hispaniola, the island that is now Haiti and the Dominican Republic, with 120 men for exploration (Fig. 11.1). The Taino people living there cultivated corn, sweet potatoes, hot peppers, yucca or manioc (a nutritious root/tuber), cotton, and tobacco. They collected traces of gold and worked it into ornaments for themselves; they did not trade it, and they had no iron. They were gentle and peaceloving; other groups on nearby islands practiced warfare. The Tainos welcomed Columbus cautiously and directed him elsewhere for gold. He left forty men, and they raped and fought so egregiously that the Tainos killed them.

On his second trip Columbus brought 1,200 men (no women) and cattle, pigs, and goats to set up a permanent colony. The men's brutal behavior—again rape, stealing gold ornaments and food—provoked the Tainos to war. After a year of fighting, during which the Spanish killed tens of thousands out of a population of perhaps 250,000, the survivors were forced to pay tribute of food, spun cotton, and gold. There were no large deposits of gold on the island; the gold in ornaments had been gathered from dust over generations, but the Spanish killed those who did not provide their allotment by chopping off their hands. The animals brought by the Spanish devoured the Tainos' food and crops, causing famine. Columbus returned two more times, always believing he was on islands off the coast of Asia, always seeking gold and spices to justify his trip. He proved a brilliant navigator but an ineffective administrator, even by Spanish standards. He returned in semidisgrace to Spain at age fifty-three in November 1504 and died in obscurity eighteen months later.[8]

Two years after Columbus reached the western hemisphere, Spain and Portugal divided the world between them—using the Spanish Borgia pope, Alexander VI, as their authority—by drawing an imaginary line down the middle of the Atlantic Ocean extending, from their point of view, around the back side of Earth. Portugal got everything to the east of the line, Spain everything to the west. They called this the Treaty of Tordesillas, hoping to prevent disputes between them (Fig. 11.2). Not yet knowing the size of the world, the treaty makers could not assign the Molucca Islands, the source of valuable spices in the East Indies. However, in 1529, after Magellan's ships had returned in 1522, Spain acknowledged that the Moluccas belonged to the Portuguese, who had already taken Malacca, the central port city on the Malay peninsula; Spain retained the Philippines.

The Spanish in the Americas acted to conquer, control, and convert nonbelievers in order to serve God and to get rich. After subduing the islands of Hispaniola and Cuba, the Spanish men looked westward for

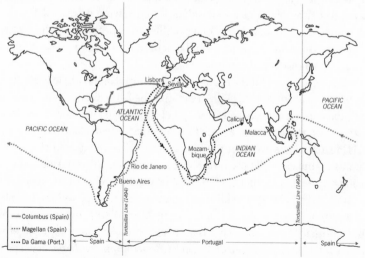

11.2 The World Divided by Spain and Portugal

greater treasure. In 1519 a thirty-four-year-old, ruthless, ambitious noble-
man, Hernando Cortez, left Cuba with 600 fighting men to assault the
Aztec empire that he had learned of two years earlier. One member of his
expedition carried smallpox; it had appeared for the first time on Hispaniola in 1518 from illicit imports of African slaves.

Cortez had no authorization either to conquer or to colonize from his
sovereign, Charles I of Spain, who had just become Charles V, Holy Roman
Emperor, as well. The Spanish king, the most powerful sovereign in Europe,
would be occupied for the next ten years trying to unite Europe and push
back the Ottoman Muslims, whom he defeated at Vienna in 1529.

The Aztecs had received reports of white-faced, bearded men for several years before Cortez's arrival. The Aztec ruler, Moctezuma Xocoytzin,
may have hypothesized that Quetzalcoatl, the god of agriculture and the
arts, must be returning, as stories had predicted he would. When Cortez
landed at Veracruz in August 1519, Moctezuma sent divine regalia, and
Cortez donned it, asking, "Is this all?" In November Moctezuma welcomed
Cortez to Tenochtitlan and put him and his soldiers up in the royal palace.
The Spanish explored the city and cleaned out the temples of human blood
in horror. When Cortez took Moctezuma hostage, Moctezuma gave him
the contents of the palace treasure room, which the Spanish melted down.
Full-scale battle ensued, Moctezuma died (exactly how is not known), and
Cortez withdrew to organize allies among those whom the Aztecs had sub-
jugated. The first epidemic of smallpox hit Tenochtitlan in 1520, wiping out
more people than the fighting had. Cortez returned to conduct a ninety-
three-day siege of the city, blockading it and leaving it without food or
water until it surrendered, with only one-fifth of its people alive. During the
siege fifty-three Spanish soldiers and four of their horses were sacrificed to
Huitzilopochtli, the god of war.[9]

It took ten years for the Spanish to gain control of the whole of Mexico,
which they called "New Spain." A year after the fall of Tenochtitlan, Cortez
became Captain General and Governor of New Spain and master of a huge
estate, with thousands of Aztecs forced to cultivate his land. He enjoyed
this life for twenty-five years until his death in 1547. By the mid 1550s, 130
Spanish families in the basin of Mexico controlled 180,000 Amerindians
using a system of semifeudal forced labor, expressly against the wishes of
Charles V. The change in culture changed the landscape forever. Forests
were cut for firewood and for building Mexico City on the ruins of
Tenochtitlan. Plows cut deeper than planting sticks and caused soil erosion. Cattle, pigs, and sheep denuded the vegetation. The Aztec canal system was abandoned. In a few generations much of the basin of Mexico

became useless for large-scale agriculture, and food had to be brought in. Spanish rule continued for nearly 300 years, until Mexicans gained independence in 1821.[10]

Smallpox reached the other American empire, the Inca in Peru, before the Spanish did. By the end of the 1520s countless Inca had died, including the emperor Huayna Capac and most of his court around 1526, and soon afterward his designated heir, Ninan Cuyuchi. The Inca had not yet heard of the Spanish; no news reached them until Francisco Pizarro landed on the Peruvian coast in 1527.[11]

Pizarro had come to the Americas to seek his fortune in 1502 as a twenty-five-year-old. He participated in conquering Hispaniola and in an expedition across Panama, where he became one of the wealthiest landowners. With a license from the king of Spain and a financial partner, Diego de Almagro, Pizarro gambled his fortune to explore the Pacific coast, where he learned of the Inca empire once he landed.

Most of the men with Pizarro were twenty-year-olds; in the Spanish feudal structure ambitious men could rise only by marrying an heiress or by warfare. Francisco Pizarro himself, the illegitimate son of a military captain who fought many battles against the Moors, was illiterate, driven to wealth and power, but not a religious zealot like Cortez. There was significant motivation, therefore, for these men to engage in violent warfare to improve their stations in life.

After the long, quixotic Spanish search for gold, Pizarro and his 180 men finally found it in the masterpieces of the Inca and in the mountains of the Andes. When Pizarro arrived on the coast of Ecuador, the Inca were embroiled in a civil war, as two half-brothers, Atahualpa and Huascar, fought each other for the throne. Atahualpa had just captured Huascar and was resting at a hot spring in Cajamarca, about 600 miles north of present Lima, before taking charge of the empire.

At their meeting in Cajamarca, Pizarro and his men captured Atahualpa after killing 7,000 to 8,000 unarmed Incas. After accepting Atahualpa's ransom of 13,420 pounds of gold objects, Pizarro executed him, set up a puppet Inca ruler, and within three years controlled the area of the Inca empire.

Pizarro and his financial partner, Almagro, argued about who would govern what territory, and in 1541 Almagro's supporters killed Pizarro. New officials had to be sent to restore order, as Spanish colonists were streaming into Peru, stimulated by a bestselling account of Pizarro's success in finding gold, which was published in Seville only nine months after the execution of Atahualpa.

In 1545 the Spanish discovered silver deposits at Potosí (Bolivia). Ten years later they found mercury in Peru, helpful for mining gold and silver, and production soared. The mineral wealth of the Andes incited further conquests by the Spanish and financed its European empire, where it was used for coins, church and palace decoration, paying debts, and expanding the army. By 1570 to 1572 the native people were resettled out of their traditional communities into new villages near Spanish centers; the overall population had declined by 50 percent and in some coastal villages by 90 percent.

Why were the Spanish conquistadors able to conquer the empires of the Americas so quickly with so few men? Why did the encounter between cultures isolated from each other for over 15,000 years result in one dominating the other with such swiftness? This drama of human history, recent enough to feel directly connected to us, haunts our imagination.

The answer seems to lie in the fact that the Spanish had the advantage of living on the Afro-Eurasian continents where people had gotten a head start on cultural specialization and invention. This happened because domesticatable plants and animals were much more abundant than in the American hemisphere, and agricultural techniques could be spread laterally in similar climates. The surplus food enabled complex societies to develop earlier, producing the skills and attributes that made the difference: guns, horses, swords, cannon, ships, immunity to disease, literacy for communication, centralized political organization for resources, and scholarship for maps and navigation. The Spanish had enjoyed the benefits of exchange with all the complex societies of Afro-Eurasia that had risen and fallen since Sumerians made the transition to urban life.

By assimilating and adapting all that Afro-Eurasians had created, the Spanish embodied strengths not realizable in the Americas, where crops could not spread easily north and south, where there were no pack animals except llamas in the Andes, and where metal technology was just coming into use. There the development of complex societies remained 2,000 to 4,000 years behind that of northern Africa and Eurasia.[12]

Of all the advantages that accrued to the Spanish and Portuguese, the one that many historians believe made the most difference was their relative immunity to the human diseases that came from animals: measles, smallpox, influenza, diphtheria, bubonic plague, and, from tropical Africa, malaria and yellow fever. Native Americans, with no exposure to domestic animals, had never before encountered these micro-organisms and succumbed in overwhelming numbers without being able to fight. The smallpox epidemic of the Columbian Exchange proved one of the two worst population disasters in recorded history; the other was the plague pandemic in the fourteenth century. At least half, and perhaps as much as 90 percent, of the Amerindian population was lost between 1492 and 1650 to repeated epidemics. The Indians encountered by the Europeans were often the traumatized survivors of intricate civilizations collapsing suddenly from overwhelming disease.[13]

The behavior of the conquistadors in the Americas did not go unopposed by theologians and the king of Spain. From 1494 on, when the pope divided the world between the Spanish and the Portuguese, scholars debated whether this gave the authority only for religious proselytizing, or also for invasion and conquest.

By 1512 to 1514 theologians in Spain were denouncing the settlers of Hispaniola and arguing that the king only had the right to proselytize, not to invade. The most famous champion of the Indians, the priest Bartolomé de Las Casas, came to their defense in 1514, and in 1520 Charles V abolished *encomienda* (giving allotments of natives to Spaniards), but he could not enforce his ruling. Twenty years later the king appointed a council to formulate new laws in favor of the natives, which provoked a civil war in Peru won by the colonists. Some experiments were made using natives as administrators, but Spaniards failed to surrender sufficient power.[14]

The Global Exchange

In the 200 years following the Spanish and Portuguese conquest of the Americas, Europeans developed the capitalist economy of the modern world. They did this by accumulating capital from both the forced labor and the land and resources of the Americas, especially the enormous quantities of gold and silver found in the Andes. With these metals, Europeans created portable wealth, which gradually overshadowed the landed wealth of the aristocrats, an irony since it was landed aristocrats who had gone out to locate the gold and silver, thinking it would increase their wealth.[15]

From 1500 to 1650 at least 180 to 200 tons of gold, worth about $2.8 billion today, flowed into Spain, carried by men out of the mountains and by mule across the Isthmus of Panama to waiting ships. Since the Spanish ruler, also the Holy Roman Emperor, used much of the gold to pay his debts, it soon flowed all over Europe, producing among other things the baroque and rococo styles of ostentatious decoration.

Yet silver unexpectedly had a greater impact, since silver coins served better than gold ones for daily transactions. With silver coins, private citizens could begin to save and accumulate wealth. In the first fifty years after the Spanish found the mountain of silver near Potosi, 16,000 tons of silver, worth $3.3 billion today, officially flowed into Europe (also carried out by men and mules), plus another possible 5,000 tons that flowed illegally. Amerindians mined the silver through forced labor, with four out of five miners dying in the first decades. From 1500 to 1600 the supply of gold and silver in Europe increased eightfold, causing inflation, which eroded the wealth of other societies. The silver coins of the Ottoman Empire lost value, doing much to undermine Islamic power. Africa also suffered, no longer having a market for its gold. India and China gained through Europeans' desire for their goods; in general, silver flowed toward China, perhaps as much as two-thirds of the global production.

In the emerging global trade, African societies had the disadvantage; all they produced that the world wanted was slaves. Portuguese sailors had first bought West Africans in 1441, taking them to Portugal and to the Madeira and the Canary Islands, where they used them to raise sugarcane. Sub-Saharan Africans had been exported as slaves to the Middle East and China for over 1,000 years, but the slave trade increased in size and significance after the Atlantic Europeans connected the Atlantic coasts. Then, with no market for its gold, African princes and chiefs sold other Africans in order to buy cloth, iron, copper, tobacco, alcohol, guns, and cowrie shells from the Indian Ocean, which were used in West Africa as the main currency. With more wealth, chiefs could acquire additional wives, who brought them more children—their chief treasure.

Europeans needed slaves in tropical America in order to raise sugar and tobacco profitably. Columbus and his men took sugarcane, originally from India, to the Caribbean, and by 1520 there were sixty sugar mills on St. Thomas island alone. But the Tainos and Caribs were dying; Africans were cheaper than Europeans, and they were resistant to malaria, which they brought with them (along with hookworm and yellow fever).

The French and English competed with Spain in the Caribbean, and by the end of the eighteenth century both the French and the English regarded the Sugar Islands as their chief trading wealth. Dutch investors competed in Brazil; by the early seventeenth century they were making profit rates of 56 percent on sugar. Annual per-person sugar consumption rose in Europe from 4 pounds (nearly 2 kilograms) in 1700 to 18 pounds (8 kilograms) in the early 1800s, providing cheap calories for industrial labor.[16]

Transoceanic slave voyages began in 1534 from Senegal and Ghana to Brazil; eventually slave export extended down to Angola and, in the eighteenth century, to Mozambique and the eastern coast. Europeans usually did not have to capture their own African slaves; African rulers and slave merchants were willing to do it. During the 350 years of the Atlantic slave trade, an estimated 12 to 25 million enslaved people were shipped from Africa for the Americas, of whom some 85 percent survived the terrible six-to-ten-week voyage. About 40 percent went to Brazil, 40 percent to the Caribbean, 5 percent to what would become the United States, and the balance to the rest of Spanish America. By the 1820s five times more Africans than Europeans had come to the Americas.[17]

During the 350 years of the Atlantic slave trade, Muslim slave traders took an estimated 2.1 million enslaved Africans from the east coast to Arabian and Indian ports. Overall, the Muslim slave trade lasted twelve centuries and shipped approximately 14 to 15 million people. Possibly as many people were enslaved within Africa as were shipped west and east.[18]

The impact of the slave trade on African culture is much debated, but at the very least it militarized many African societies, promoting warlord-entrepeneurs. In terms of loss of people, the impact of slavery on Africa proved far less than that of disease on the Americas.

The English government got into the race for colonies early by sponsoring a Genoese sailor, John Cabot (Giovanni Cabato), who reached Newfoundland in 1497 and New England in 1498. After these trips large-scale fishing expeditions were launched to harvest the bounty of the waters off North America, which required salting to reach Europe in edible condition. Salt cod fed sailors, soldiers, and the poor of northern Europe. England was not quick to follow up on Cabot's landing. Sir Walter Raleigh tried and failed on an island off North Carolina in 1584; English colonists barely succeeded at Jamestown in 1607 and Plymouth in 1620. The English government provided royal charters but was not able to provide strong bureaucratic support; its colonists were more nearly on their own than were Spanish or French colonists.

The French experience in the New World focused on furs, for which there seemed to be insatiable markets in China and Europe. The French founded a colony, called New France, on the St. Lawrence River in 1608, at the site of present-day Quebec. From there French trappers moved up the St. Lawrence River and out onto the Great Lakes and Hudson Bay to penetrate the heart of the continent to secure foxes, ermine, squirrels, and sable, easy to transport in canoes. In exchange for fur, Amerindians received firearms, fabrics, metal tools, and alcohol. With the trappers went Jesuit missionaries full of zeal for converting natives to Catholicism. Not nearly as many colonists settled as did in the English colonies; French royal policy excluded Huguenots (French Protestants) from going to America because they wanted New France to remain Catholic. (The Reformation had begun in 1517; see page 206). After a series of wars between France and England and their colonies, the French in 1763 had to cede Louisiana to Spain and Canada to England.

While the French were bringing the hunting-gathering peoples of North America into the trading network of agrarian and urban societies, the Russians were doing the same in the vast area of Siberia, roughly one quarter of Eurasia. There people still lived in kinship groups as hunters, fishermen, gatherers, and herders of reindeer/caribou. The Russians recruited Cossacks from the Ukraine to build a fur business in Siberia, using cannon on riverboats to defeat the local people; the diseases they brought helped them, too. By 1440 they reached the shores of the Pacific, imposed a tribute system on every adult male, and traded with flour, tools, and alcohol in exchange for furs. During most of the seventeenth century the Kremlin received 7 to 10 percent of its revenues from the pelts arriving from Siberia. By the 1730s Russians had extended their trading posts to Alaska, and by 1810 to northern California.

One other European country, the Netherlands, competed for colonies in North America. It set up a trading colony at New Amsterdam, at the mouth of the Hudson River, and pushed into the fur trade. The Dutch were stopped in 1664, when the English navy took New Amsterdam by cannon and renamed it New York.

Out of this trading frenzy in the Atlantic Ocean emerged new economic institutions that would reshape the world. Spain and Portugal in the fifteenth and sixteenth centuries tried to keep their overseas trade and colonies as royal monopolies, but this proved expensive and inefficient. Colonists in America preferred to trade with the French, Dutch, and English. Wealthy private investors from those countries figured out ways to reduce their risk and increase their profits through banks, stock companies, stock exchanges, and chartered trading companies, which together formed the system called capitalism. Joint stock companies sold shares to investors as a way to raise the large sums required for overseas expeditions; investors needed a way to buy and sell shares. In 1530 a stock market opened in Amsterdam that became the world's largest through the seventeenth and eighteenth centuries; in England the Royal Exchange, from 1695, developed into the Stock Exchange in 1773.[19]

Some governments supported the efforts of their private citizens with policies to promote and defend their citizens' trade, with force when necessary, a policy called merchantilism. One of the early examples of merchantilist capitalism occurred in the Netherlands, when the government in 1602 gave the Dutch East India Company a legal monopoly over all Dutch trade in the Indian Ocean. This encouraged investors to buy shares in the company; they were rewarded when the company captured control of the sea routes in the Indian Ocean from the Portuguese. The government was rewarded with increased taxes. About twenty years later the Dutch government chartered the Dutch West Indies Company to operate in the Atlantic, where it seized Brazilian ports and African slaving ports from the Portuguese. From 1652 to 1678 the English and French defeated the Netherlands in the Americas using larger navies.

As economic institutions developed, banks charged interest on money lent, called usury by those opposed to interest or at least excessive interest. The demands of the emerging capitalist economy created a crisis of conscience for those Christians who, like Muslims, believed that charging any interest was wrong—an agrarian reaction to the power of money. The Old Testament (Deuteronomy) says that one may charge interest to strangers only. Catholics in Europe reasserted ancient restrictions on lending money at interest; it continued to be an offense in French law until 1789. Calvin stated the Protestant position in 1545; that interest is legitimate at reasonable rates of about 5 percent.[20]

The trade between Europe and Africa and the Americas involved two kinds—items already known to both sides and items entirely new to one side. For example, fish and fur were known to both sides, but sugar had never been seen in the Americas, while tobacco had never been grown in Afro-Eurasia. Historians call this transfer of novel peoples, plants, animals, diseases, and technology the "Columbian Exchange."

It is difficult to imagine culinary life before the Columbian Exchange. What did Italians put on their pasta before tomatoes came from America? Likewise, chocolate came from the Americas; no one in Afro-Eurasia had any before 1492. Corn and potatoes, new to Europe, helped keep many people alive. Cassava, or manioc root, a highly caloric root crop of tropical America (native to Brazil) that thrived in poor soil and drought, became a lifesaver in tropical Africa. Beans, squash, sweet potatoes, peanuts, chilies, dyes, tobacco, and medicines all flowed to Afro-Eurasia as gifts from the Americas.

Flowing back to the Atlantic coast were the European crops—wheat, olives, grapes, and garden vegetables. Also, Europeans took the Afro-Asian crops—rice, bananas, coconuts, breadfruit, sugarcane, citrus fruit, melons, figs, onions, radishes, and salad greens. The Spanish took horses, which had originated in the Americas but had died out there during the last ice age, cows, pigs, sheep and goats, rats, and rabbits. The enslaved Africans took okra, black-eyed peas, yams, millet, sorghum, and mangoes.

Along with the gifts of Afro-Eurasians went bacteria and viruses absolutely new to the Americas. Peoples in China and the Mediterranean had been devastated in the early centuries of the Common Era but had gradually developed immunities. The Amerindian population, virgin to the diseases, was reduced by 50 to 90 percent, as described earlier. The Americas sent tobacco back to Afro-Eurasia, and it seems likely that syphilis originated in the Americas, taken by sailors to the rest of the world.

The expanding world trade eventually brought the ecosystems of Australia and the Pacific Islands into contact with the rest of the planet, though not until the late eighteenth century. In 1769 Captain James Cook, sponsored by Great Britain, began charting the coastline of New Zealand, where an estimated 100,000 Maoris lived, descendants of Polynesians who had arrived about 1300. Australia was home to some 750,000 Aborigines, who lived in mobile groups hunting and gathering. In 1788 Britain began shipping convicts, mostly petty thieves, to Australia; by 1845 the settlers outnumbered the Aborigines. In the exchange Australia gave the world the eucalyptus tree, while it acquired scores of new plants and dozens of new animals in the most drastic alteration of any area of the world.

The Major Empires

Despite the exchange going on across the Atlantic Ocean and on its coasts, China and the Moghuls in India remained the world's powerhouse empires in volume of trade and wealth through the eighteenth century. After China and India, the two other most powerful governments were the Ottoman Empire, originating in Turkey, and the house of Hapsburg, controlling about 20 percent of Europe and the Spanish colonies of the Americas.

In China the Ming dynasty passed to the Qing, a Manchurian family that captured Beijing in 1644, conquered the rest of China in forty years, and ruled until 1911. The Qing dynasty included two outstanding emperors, Kangzi (r. 1662–1722) and Qianlong (r. 1736–1796). The Chinese government controlled trade strictly, allowing Europeans to trade only at Canton, where they exchanged mostly silver for the Chinese products they wanted. The wealthy and aspiring middle classes of Europe avidly consumed Chinese goods, genuine and imitation—silk, porcelain, tea, wallpaper; by the late 1700s the British were worried about their massive trade deficit with China and tried to negotiate changes in policy, to no avail.

The Chinese population shot up from 100 million under the Ming to 350 million by 1800, constituting one-third of humanity. The pressure of population led to deforestation, which led to serious erosion and flooding; the Grand Canal was so silted as to be unusable by the end of the eighteenth century. Rebellions by farmers became endemic in central and southwestern China.

The Moghul empire in India developed when Muslim Turks from the Ferghana Valley in today's Uzbekistan, led by Babur, defeated the sultanate of Delhi in 1526. Babur was a descendant of Genghis Khan's second son, Chaghatai. These Muslim Turks ruled as a minority over a land of Hindus and extended their empire over all of northern India from the Indus and Ganges rivers, Kashmir, the Punjab, and down to Bombay, but not the southern tip or eastern coast. Babur's grandson, Akbar (r. 1556–1605) married a Rajput princess, producing a son both Muslim and Hindu. The court produced a distinctive style based on sumptuous display, incalculable wealth, and the Persian language. With a population of 100 to 150 million, India enjoyed great prosperity in the late sixteenth and early seventeenth centuries, trading cotton cloth from its ports, with no navy or merchant ships. In the seventeenth century, Moghul state revenues were four times those of France. The Moghul empire survived in name until 1857, when the British removed the last Moghul emperor, but its real power began to evaporate after 1707 as regional Hindu power challenged Moghul military supremacy, devolving India into princely states and leaving them vulnerable to British intrusion in the nineteenth century.

The Ottoman Empire, formed by Muslim Turks from northwestern Anatolia, started growing after 1415, included Constantinople (changed to Istanbul) by 1453, and by 1550 extended from the Euphrates River to Hungary in Europe and the Sahara in Africa. The empire contained some 20 to 25 million people, growing perhaps to 30 million in the eighteenth century. After conquering Christian countries in the Balkans, the Ottomans imposed levies of male children on Christian villages, placing these children with Turkish families and later in military schools in Istanbul to provide soldiers and, for the talented few, government posts. Coffee, grown in the highlands of Yemen on the tip of Arabia where the Red Sea meets the Indian Ocean, became the rage in fifteenth-century Istanbul before it spread to Europe (Venice in 1615, London in 1651). Despite frequent wars with Iran, the Ottoman Empire lasted until the tumult of World War I put it out of business.

In 1600 Iran, known as the Safavid empire, stretched from Baghdad in the west to the Moghul empire in the east. The founder of its ruling family, Ishmael, insisted that Iran practice Shiite Islam, distinguishing it from its neighbors, all of whom were Sunni. Ever since the Mongols destroyed Baghdad in 1258, Iranian culture had looked more to India than to Arabia; its scholars and writers used Persian primarily, rather than Arabic. Most people in the Safavid empire, whether Iranians, Turks, Kurds, or Arabs, lived by subsistence farming or herding. Their only products for foreign trade were silk fabrics and the deep pile rugs still treasured today. By 1722 the central government had so little support from its nomadic groups that Afghans were able to capture the capital and end Safavid rule.

The fourth of the world's largest empires from 1000 to 1500 was the house of Hapsburg in Europe; it originated in Switzerland, where the ruler, Rudolf I, became Holy Roman emperor in 1273. He installed his son as ruler of Austria, and through marriage and inheritance the Hapsburgs acquired the Netherlands in 1477 and Spain in 1516, plus Luxembourg, Burgundy, Bohemia, Hungary, Sicily, Naples, and Milan. In 1519 the Hapsburg king of Spain, Charles I (the grandson of Ferdinand and Isabella) became, as well, the Holy Roman emperor known as Charles V, as mentioned earlier.

Since Charles V ascended the throne just two years after Martin Luther started the Protestant Reformation, he led the militant Catholic crusade against Protestantism, as well as the crusade against the Muslims of the Ottoman Empire. Charles V ruled over 20 million people, about 20 percent of Europe. He was not successful in unifying Europe or in stopping Protestantism; he fought four wars with France, and the Holy Roman Empire degenerated into religious wars between Protestants and Catholics. Charles V abdicated and retired to a monastery in 1556; the holdings of the Hapsburgs were broken up, continuing in Austria to the end of World War I in 1918.

By the early 1600s Spain was bankrupt, despite its subsidy of silver from its American colonies, and France, led by the Bourbon kings, had become Europe's most powerful state. Amsterdam served as the financial center and major port of seventeenth century Europe. By 1689 England had become France's most powerful rival. The political fragmentation in Europe created an unusually competitive situation and possibly a self-generating process of technological change.

Rural life in Europe probably worsened from 1500 to 1750. Cooler weather, known as the Little Ice Age, caused illness, malnutrition, and death. Trees were cut in immense numbers for lumber for ships and buildings, fuel for heating and cooking, and charcoal for smelting ores. This deforestation affected the poor, who flocked to cities. In 1500 Paris was the only northern European city with over 100,000 people; by 1700 Paris and London each had 500,000 people, Amsterdam had 200,000, and twenty others had more than 60,000. In the cities, the wealthy people who dominated, called the bourgeoisie or town dwellers by the French, devoted long hours to business, supported by monarchs whose state revenues increased as business grew. Some 10 to 20 percent of urban people were so poor that they were exempt from taxation.

Religion, Science, and Warfare

From 1450 to 1800 printing with moveable and reusable metal type distinguished Europe and its colonies from the rest of the world, including the Ottomans, Moghuls, and Ming Chinese. These other empires relied on scribes until the nineteenth century. The possible reasons are varied and unknowable—perhaps governments feared they couldn't control printing or feared offending the scribes; perhaps printing didn't seem much better in places using ideographic rather than alphabetic scripts. For a time, Muslim authorities felt that printing constituted a desecration of the sacred text of the Quran. The only known exception was Korea, where moveable metal type had been invented in the thirteenth century and an alphabet added in the early fifteenth, setting off a boom in printing restricted to Korea, where only a tiny elite could read.[21]

In Europe Gutenberg's printing press (1454) spread with amazing rapidity. By 1500, 236 European towns had similar presses. By 1501, type was cast for the Cyrillic and the Greek alphabets. The Spanish set up presses in America by 1533 and the English in 1639. By 1605, the first regular newspapers appeared, and by 1702, the first daily. By 1753, in Britain, 20,000 newspapers were sold daily. Rates of literacy rose, business communication flowed, and more people took part in intellectual debate, particularly in vicious religious arguments.

It was probably printing that led to nearly half of Europe becoming Protestant. A Catholic monk and professor of divinity at the University of Wittenberg, Martin Luther (1483–1546) objected to Pope Leo X's practice of authorizing indulgences, by which people could earn forgiveness of the punishment due for past sins by making a donation or going on a pilgrimage. Luther wrote his objections in Latin and nailed them to the church door in 1517, asserting that Christianity is a personal commitment and that salvation comes by belief/faith alone, to the exclusion of any kind of good works. Cheap pamphlets spread these ideas, which were embraced by princes and paupers alike; the established church tried but could not stop their spread. Luther didn't argue for religious toleration; he wanted Protestantism to replace Catholicism as the one true faith. The Germanic states dissolved into warfare, and in 1555 a truce was negotiated in which every prince was permitted to decide the religion of his own kingdom. In the rest of Europe, other reform movements sprang up, such as Calvinism, Anglicanism, and Presbyterianism.

Challenges to established religion occurred in many areas of the world. In China Wang Yangman (1472–1529) argued that ordinary people could achieve virtue and truth without long studies in Confucian ideas; his challenge eventually resulted in a reassertion of Confucian orthodoxy. In India, Gure Nanak (1469–1539) started a new religion, Sikhism, based on Hindu texts but rejecting the clerical authority of the Brahmin caste and advocating a strict moral code for all followers rather than one that varied for different castes as did Hinduism. The emperor Akbar (r. 1556–1605) encouraged religious diversity and toleration, but later emperors favored Muslims.

Around the world during this time many local religions disappeared, swept away by the increasing currents of movement, trade, communication, and colonization. People who clung tenaciously to local traditions were left isolated or became opportunities for Muslim and Christian missionaries. Even the nomadic Mongols, who had long followed local shamanistic traditions, gradually converted to Tibetan Buddhism; in 1601 a Mongol was chosen as the next Dalai Lama and soon Buddhist monasteries sprang up all over Mongolia.

In Europe the challenge to authority went to the point of forming a new attitude: Don't rely on authority at all but subject every idea to experimentation and reason. This attitude, known as scientific, flourished in Europe, where universities served as supportive communities for those practicing science. By 1500 more than one hundred universities functioned in Europe, absorbing and trying to make sense of the flow of global information coming their way. The papacy began in 1559 to proscribe books it considered subversive; it did not give up trying until 1966. It acted against the scientist Galileo Galilei, who argued that the Earth circled around the sun rather than vice versa. Church authorities kept him under house arrest in Florence, Italy, after 1616, but they could not suppress his ideas, which were published in the Netherlands. The conflicts produced by the rapid changes going on in the world were reflected in the plays of the Englishman William Shakespeare (1564–1616).

Warfare has been a constant thread in this narrative without much description of how it was conducted. From 1450 to 1800 the nature of warfare was changed by major new developments: oceanic navies armed with cannon, field cannon, and huge fortifications against them; standing drilled armies capable of not retreating under fire; and massive logistical support for field troops. These developments cost so much money that banking systems became part of warfare. The largest empires spent 70 to 90 percent of their revenues on their war machines.

Not every empire engaged the new military developments equally. Only Europe and China built oceanic navies, and China retracted its. The Europeans and Ottomans excelled in field cannon and armed infantry. Europeans invented close-order drill plus the banking system (Italy, the low countries, England), despite the fact that Christian scripture prohibited usury. Muslims obeyed their scripture and practiced no usury; they figured out a system whereby the loaner became a partner in the venture. The Moghuls had no banking or sea power. The Qing defeated the Ming in China by using siege cannon, which they learned to use from Jesuit missionaries. Africans acquired only a small part of the new military developments, which destroyed nomad power forever since nomads could not make guns and cannon in quantity.

By 1750 to 1800 a worldwide system of exchange and trade was in place, using the seas that connected the continents. Port cities and their hinterlands prospered, while landlocked areas languished. It took a month to cross the Atlantic, three months to cross the Pacific, a month or more to cross the Sahara by camel, and a year to walk from one end of Eurasia to the other. Trade on the high seas distinguished the period from 1450 to 1800.

During the same period the world's population doubled to reach 900 million. The Columbian Exchange of food crops contributed to this, as well as the waning of disease epidemics. Africa grew in population much slower than other areas, but in general the eighteenth century proved the turning point in population growth; it marked the beginning of the modern age of extremely rapid growth. Yet 80 to 85 percent of the world's people still lived as farmers, using their own muscles as power, not being able to read, and only seldom encountering a stranger. The number of slaves in the world grew markedly during this period, reaching 20 to 50 million, or 2 to 5 percent of humanity by 1800.

The growth of the global economy proceeded at about the same rate as the population. From 1450 to 1800 the economy grew less than a quarter of a percent annually, for a total increase during this time of two- to threefold. The growth occurred mostly from population increase rather than from increased efficiency, since muscles still provided the power for most efforts. China and India remained at the economic center with almost 80 percent of the world's goods and services until the mid–eighteenth century, when the Atlantic economy began to rival that of the western Pacific. Within a few decades after 1492 the rate of economic growth in Europe began to speed up dramatically, based on cheap and slave labor and land taken in the Americas.

For the period 1000 to 1500 our story focused on the Mongols. For the period 1500 to 1800 it has focused on the Atlantic Europeans, the Mongols of the sea.[22] With ruthless brutality and determined conviction, they extended their culture to the Americas, and in Europe they laid the foundation for industrialization and the worldwide dominance of Europe that followed.

Unanswered Questions

1. What impact did the slave trade have on Africa?

Some historians, such as McNeill and McNeill, judge that the impact of the slave trade on the demographics of Africa as a whole was probably small. They say that 25 million enslaved people, spread over 400 years and many countries, affected a very small (but unknowable) portion of the population.[23] Others, such as Patrick Manning, say that sub-Saharan Africa had virtually no population growth from 1750 to 1850 and that without slavery it would have grown from 50 million to 70 or 100 million. Manning believes that from 1750 to 1850 "perhaps 10 percent of the African population—6 or 7 million people—was in slave status as a result of Occidental and Oriental demand for slaves."[24]

Slavery had other impacts on Africa. It favored the formation of states because stateless people were especially vulnerable to capture. It militarized many societies and increased the number of guns, purchased from trading slaves. It divided societies and forced societies and individuals to choose whether or not they would trade in slaves. Even today in parts of Africa people remember whose ancestors were slavers and whose were slaves.

2. Did any Europeans protest the treatment of the Amerindians?

A few did, chief among them Bartolomé de Las Casas (ca.1484–1566), born in Seville of converso heritage—that is, his family converted from Jewish to Catholic faith. His grandfather was burned as a Jew in 1492; his father accompanied Columbus on his second trip. Las Casas went with his father to the Caribbean from 1502 to 1506, returned to Spain to take the orders, and returned to the Caribbean in 1509 to 1515 to witness the conquest of Cuba. Las Casas wrote several books detailing the mistreatment of the native people at the hands of his countrymen. Eventually he influenced the passage of laws that offered some protection to Amerindians, called the New Laws of 1542, which outlawed the enslavement of Amerindians and limited other forms of forced labor. In the 1550s he wrote *A History of the Indians*. Las Casas, however, did not object to the enslavement of Africans.[25]

3. Why did capitalism emerge in parts of Europe rather than in parts of China or India?

This is a favorite question of social scientists and historians in Europe and the United States. Many possible explanations have been advanced, including the competition among states due to the lack of a European unified government, the location on the sea, the windfall of food, labor, and gold and silver from the conquest of the Americas, the strong state machineries, deposits of coal, population density, fast communication due to printing, differences in social structure and policy, plus lots of accidents. All seem to have played a role. The follow-up question is: How did England come to take the lead in the development of capitalism? The next chapter will offer some suggestions.

12

Industrialization

(1750–2000 CE)

To describe the change to fossil fuels, to the factory system, and to a manufacturing economy, which began to occur first in England about 1750, some historians use the term "industrial revolution." Others, including myself, prefer to use "industrialization" to indicate a much longer, gradual process that began as Afro-Eurasia became one network (chapter 10), intensified with the connection of the globe (chapter 11), and completed itself in England about the 1850s, later in other places, and is still going on in many places. Historians mostly agree that the human switch to fossil energy and manufacturing represents one of the three or four fundamental shifts in human history, as significant as the shift to agriculture or to cities.[1]

In the development of industrialization, two phenomena new in human history occurred. One concerned population growth, the other economic growth.

From 1 to 1700 CE the world's population grew gradually, averaging about 12 percent a century. But growth was not continuous; periods of falling population occurred, when the number of people outstripped food or disease decimated society, as noted earlier. After 1700 the death rate began to fall, and the world's population grew 30 to 50 percent in the eighteenth century, 80 percent in the nineteenth century, and 280 percent in the twentieth century. No one knows why—warmer climate, more food (from new varieties from the Americas and from better agricultural techniques), better transportation, and more circulation of lethal disease (hence more immunity) seem the most likely explanations (Fig. 12.1).[2]

Even more dramatic, from the late 1600s the English and Dutch

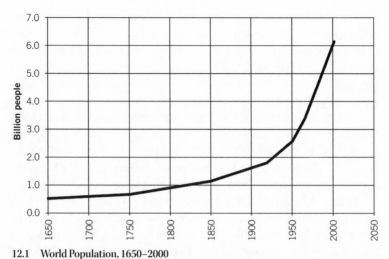

12.1 World Population, 1650–2000
(*Source:* Donella Meadows, Jorgen Randers, and Dennis Meadows, 2004, *Limits to Growth: The 30-Year Update,* White River Junction, VT: Chelsea Green Publishing, 6.)

economies increased the per capita income of their growing populations, despite diminishing returns in agriculture. Per capita income growth (a rise in real income per person) had occurred before in various parts of the world, but it had always been reversed, leaving the standard of living of subsistence farmers about the same as it was before. In Europe, however, the standard of living has not been reversed, except temporarily in time of war, since the late 1600s. No one knows the causes for this, although much analysis has been made, the best known being that of Adam Smith who argued, in *The Wealth of Nations* (1776), that societies can promote economic growth by establishing peace, low taxes, and impartially implemented laws to protect property and investments.[3]

Bourgeois Power

Since the establishment of the first states about 5,000 years ago, the most common and durable political arrangement has been the monarchy—rule by a single person, with variable constraints on his (or occasionally her) power. Only a few societies, since the time that cities emerged, have created democracies in which every citizen, however defined, took part, and those societies remained small scaled and short lived.

Once trade increased and the number of cities grew dramatically, beginning about the seventeenth century, powerful communities of merchants and commercial landowners resented the taxes levied on them by monarchs. Large empires, based on ancient traditions, like the Qings in China, the Moghuls in India, and the Ottomans in Turkey and the Middle East, managed these tensions, avoiding civil war.

In Europe, however, where competing small, young states prevailed, the monarchs in two states—Holland and England, could not manage the tensions. Bourgeois leaders took their societies into civil war to limit the power of their monarchs. In Holland the urban elite cast off the Hapsburg authority after a long war from 1567 to 1609 and set up the Dutch Republic.

In England the political revolution, coming fifty years later, also took about fifty years to complete, from the outbreak of civil war in 1642 and the beheading of Charles I in 1649 to the restoration of his son, Charles II, in 1660, and finally to a coup d'etat by Parliament in 1688 to 1689. Men of property in Parliament carried out this political revolution; they arranged a monarch as a symbol of unity, with real power situated in Parliament, as stated in a bill of rights, which guaranteed Parliamentary supremacy by control of the purse, regular and frequent meetings, and prohibition of power to the Crown. Known as the Glorious Revolution, this political change laid the foundation for the dramatic economic and technological changes that followed. The elite who led this revolution, about 5 percent of the population, controlled about 25 percent of the national income. They centralized and reformed public finance, setting strict standards for bookkeeping.[4]

Other political revolutions arose from the increased Atlantic trade. British colonists in North America, mostly men of property, resisted higher tax levies and declared war on the British in 1776. The colonists won their independence because overseas war proved too expensive for the British and because the French lent their assistance to the colonists. In France the political revolution followed shortly, in 1789, when the peasants joined the Estates General, mostly men of property, to overthrow the monarch. The French endured a period of anarchy, followed by the dictatorship of Napoleon Bonaparte, then a restored monarch followed by more revolutions, until a lasting republic emerged after 1871. In the Americas the French sugar colony of Saint Domingue revolted and won its independence as Haiti in 1804. By 1826 an independence movement in Latin America, led by Simon Bolivar (1783–1830), divided most of Spanish holdings in America into independent nations, leaving only Cuba and Puerto Rico, where planters remained loyal to their best customer, Spain. Brazil achieved independence from Portugal in 1822. In Germany the monarchy remained, with a limited parliament, until defeat in World War I, and the Russian monarchy fell in 1917.

Bourgeois men of wealth and property developed the idea of representative government as they sought to counter the power of monarchs to tax their wealth. Two significant Englishmen who articulated the ideas of representative government were the poet John Milton (1608–1674), and the advisor to an earl, John Locke (1632–1704), both men of secure wealth. Milton, the son of a money-lender who funded his son's future, became secretary of the Commonwealth that lasted from 1649 to 1660, when Parliament ruled through its victorious general, Oliver Cromwell. During the civil war that preceded the Commonwealth, Milton wrote his ringing defense of freedom from government censorship, the *Areopagitica*. In the pamphlet war that followed the end of the king's censorship, Milton defended Parliament's right to kill the king.

A few years later John Locke, too, defended the rights of Parliament against the rights of a monarch in *Two Treatises of Government* (1690). In these essays Locke argued that men submit to government, not because the inclination to submit to an absolute leader is innately present in people, but because legitimate government protects their right to property. If government does not offer this protection, then men of wealth have the right to withdraw their consent to government and to form a new one. Kings have no right to rule, but rather men have a right to consent. To strengthen his case, Locke developed a new theory of mind, in which he denied there were any innate ideas, including the inclination to submit to absolute authority. On the contrary, Locke argued, the mind of a newborn is a *tabula rasa*, or blank slate, into which all ideas come from experience and reason.

But which people have the right to consent to government? In Locke's day, only a handful of adult males voted for members of Parliament. When Thomas Jefferson (1743–1826) used Locke's ideas in the *Declaration of Independence* (1776) to justify the revolt of England's colonies in America, he substituted in his list of rights the "pursuit of happiness" for "property," apparently broadening the base of what he considered legitimate voters. James Madison clinched the argument for Jefferson by asserting that, while property in one sense means land, money, and merchandise, in a wider sense it means free opinions and the free use of one's faculties as one chooses, concluding that "as a man is said to have a right to his property, he may equally be said to have a property in his rights."[5] With these ideas, the consequences of the actions of the wealthy men who challenged and beheaded King Charles I may have been taken further than they ever intended.

Nothing was put into the U.S. Constitution about voting requirements. The Electoral College voted for the president, the state legislatures voted for senators, and each state set its own requirements for electing men to the House of Representatives. Only Pennsylvania had no property requirement in 1800. Slaves could not vote but counted as three-fifths of a person in determining the number of congressmen, giving the slave states extra power. No woman voted until 1920.[6]

As the idea of the political equality of men gained acceptance in Europe and the United States, people in these regions developed the concept of race as a type classified by physical characteristics. If all men are equal, how can it be explained that some appear so backward? the Europeans questioned. The Swedish naturalist Carl Linnaeus made a stab at racial classification in 1735, but the most authoritative one was established by the father of physical anthropology, Johann Blumenbach, a professor at the university in Göttingen, Germany, on the basis of measurements of skulls. In his *On the Natural Varieties of Humankind* (1775), Blumenbach distinguished five races of humans; by the third edition of the book in 1795 he had named these Caucasian, Mongolian, Ethiopian, American, and Malay. Blumenbach did not believe that Africans were nearer to apes than other people, but he did believe the Caucasians were the first race from which others diverged.[7]

Industrial Revolution

The change to industrialization is now being seen as a global phenomenon, created not by European society but by the forces at work in the whole global network, the interaction of people in Afro-Eurasia with the people of the Americas. After the connection of the two hemispheres, a sharp acceleration occurred in rates of innovation, levels of productivity, and the pace of collective learning around the globe. The Atlantic seacoast of Europe benefited from its location as the first hub of the first world system; Europeans were exceptional by their strategic location and by being young and flexible, ready for change.

The process of industrialization began specifically in Britain, the damp little island off northwest Europe. One major reason lay in the fact that Britain had large deposits of coal. As the island began running out of forests to convert into charcoal for smelting, iron production began to slump. Burning regular coal for smelting iron did not work, since the impurities in the coal made the iron brittle. In 1709 the Darby family in Shropshire discovered that when coal is first converted into coke, smelting could be done successfully.

Coal deposits, however, were deep and were impeded by water filling the shafts. Some kind of pump was needed. In the 1770s a Scotsman, James Watt, improved the design of the steam engine, and by 1800 Britain had about 2,000 steam engines—still only about 5 percent efficient but each equal to some 200 men pumping water out of coal mines. (Coal output grew almost 500 percent from 1780 to 1830.) Improvements had to be made in the steam engine; for this, the English used all their gun-making and clock-making skills. Costs of energy for steam-run equipment went down precipitously after 1830.[8]

Steam engines as curiosities had developed in other societies before the eighteenth century. The Chinese had various systems, but they used the engine as a bellows in which the wheel turned the piston rather than the piston turning the wheel, as in Watt's design. The Chinese also used coal to produce iron; in 1080 their iron production exceeded that of non-Russian Europe in 1700. But northern China, where the coal deposits lay, suffered Mongol invasions, civil war, floods, and plague. Population shifted to the south, and iron production, when it recovered, used charcoal instead of coal.[9]

In the 1700s the world's only large exporter of cotton fabric was India. In 1721 wealthy merchants in England, through Parliament, prohibited the import of Indian cloth, in order to increase their own income from local production. They procured the raw cotton from American colonies, produced by slaves, and took it to rural areas of England, where whole families of artisans used hand tools—the spinning wheel and the home loom—to produce cloth, which the merchants marketed. The spinners and weavers worked at home, or in small groups like protofactories.

In 1764 a society in London offered a prize for the best improvement of the spinning process; James Hargreaves won it with his "spinning jenny," a wood frame with a series of spinning wheels geared together to produce eight threads simultaneously, or 100 threads when adapted to water power. By the early 1800s the English had figured out a loom powered by steam engine; the invention of the cotton gin in Georgia (United States), increased the output of cotton. By the 1860s India could not match British competition in cotton textiles.

The process of industrialization in England required many simultaneous changes. Inventions, as described, were necessary. The American colonies provided raw materials and markets. Canals and roads provided a basic transportation system; later, steamships and railroads speeded up transportation. Financial systems to support the accumulation of capital had started to develop in the seventeenth and early eighteenth centuries. Attitudes about usury had to change. Finally, increased agricultural production released workers from the fields to the factories.

English farmers made significant increases in their production by selective breeding of sheep to double their average weight, by planting seeds in rows instead of by sowing, by using horse-drawn drilling machines, and by figuring out a four-year crop rotation (turnips, barley, clover, wheat) that did not require leaving fields fallow. By feeding turnips to their cattle over the winter, farmers did not have to slaughter in the fall and had milk and butter year round. These changes required larger fields to be effective; the wealthiest farmers demanded enclosure of formerly common areas where poorer farmers grazed their animals. The enclosure movement peaked in the last decades of the 1700s and the early decades of the 1800s, as small farmers became hired hands or left farming for the city. Even though food production increased, so did population, and England by the mid-1800s needed to trade manufactured goods for foods. The last surplus of English wheat for export occurred in 1792.[10]

All these changes proved difficult, if not catastrophic, for the poor people of England. Thousands of hand weavers were thrown on the streets by the use of power looms. Wages fell from 1760 to 1815. Historians do not agree on whether conditions for the poor were worse than earlier or, if so, how much worse. Some say that two generations were sacrificed to create Britain's industrial base, but most agree that after 1850, when industrialization is considered mature in Britain, the people as a whole shared in Britain's success on the world stage. Meanwhile, a massive migration took place; from 1815 to 1914 20 million Brits left their island. In 1900 the British population was 41 million, but it would have been 70 million without the exodus.[11]

It is no coincidence that tobacco, cocoa, tea, and coffee became features of everyday life during the process of industrialization. Tobacco from the Americas was cultivated in England by 1565, coffee reached London in 1651, chocolate in 1657, and tea in 1660. All were addictive, quick to prepare and consume, and provided short bursts of energy—perfect for long days away from home. By adding sugar to the drinks, the poor were able to prevent the scarce protein in their diets being burned for energy. An acre of tropical sugar land yielded as many calories as four acres of potatoes or nine to twelve acres of wheat. By 1900 sugar imports had increased eleven times since 1815, and Brits averaged 15 to 25 percent of their daily calories in sugar.[12]

Since industrialization removed work from the family setting, it impacted women and children in a major way. They became a flexible workforce, entering the labor market as needed to supplement men's labor, holding jobs that men didn't want, that held little authority or required minimal training. In this way inequality was built into the process. Yet some children had the advantage of being released from productive/grinding labor to pursue education as their primary task, and some women in urban households and service industries (excluding domestic service) were better off than male day-laborers in agriculture.[13]

Why did industrialization begin in England when it did? Historians have come up with various answers. The short one seems to be a combination of unique factors: location on the sea, overclearing of forests, coal deposits, the social and political consequences of the Glorious Revolution, commercial and agricultural development based on land and wealth in the Americas, transportation, instrumentation skills, population growth, printing presses, plus the freedom and incentive to innovate.

After 1815 other parts of Europe and the United States began the process of industrialization. Belgium and Switzerland industrialized early; they had deposits of coal. Germany had the coal-rich Ruhr region, and its industry outstripped the British by the 1880s. France had too little coal to be a leader and had to import it after 1848. The United States pioneered factory management, first using interchangeable parts in weapon production, and by the 1890s its industry outpaced that of Germany to be the world's leader. The only non-Western societies that began to industrialize before 1900 were Russia and Japan. Russia began in the 1860s and by 1910 had the world's fourth or fifth largest heavy industry complex, fully mature by 1950. Japan also began in the 1860s and by 1914 had become a first-rank military and industrial power. The great powers of the twentieth century proved to be those who had managed to industrialize in the nineteenth century—Britain, Germany, Russia, the United States, and Japan.[14]

With the harnessing of coal, which made labor less scarce, slavery and forced labor gradually became less attractive or economical. Right at the height of slavery and serfdom in the world, these two ancient arrangements were, rather rapidly, mostly abolished worldwide.

The peak of slavery and serfdom came in the first half of the nineteenth century. Slavery quintupled between 1800 and 1860 in the U.S. South to produce cotton. It expanded in the Caribbean and Brazil to produce more sugar. In southeast Asia slaves on plantations produced sugar and peppers. In Russia millions of serfs raised wheat; in Egypt they formed the army and raised cotton; in North Africa slavery increased during this time, especially to raise palm oil, used as an industrial lubricant.

Agitation to abolish slavery began with the Quakers in England and with the enlightenment philosophers in France in the late eighteenth century. Printing and travel circulated the idea. By 1807 in England and 1808 to 1830 in France the selling of slaves was abolished. In the 1820s Chile and Mexico abolished slavery itself; England did so in 1833. Other Atlantic countries followed: the United States in 1865, Spain in 1886, Brazil in 1888. In 1861 Russia abolished private serfs, who had to work at least nine more years to own their land communally; government serfs were freed in 1866. The Ottomans succumbed to European pressure and banned slave trading but never slavery itself, since it was recognized in Muslim law. In Africa trading ceased by 1914, and abolition came in the first third of the twentieth century. On the whole, the abolition of slavery and serfdom represented a historic liberation for humanity; 50 million serfs in Russia alone gained their freedom. The use of fossil fuels helps explain why slavery has officially if not completely vanished.[15]

Inventions in transportation and telecommunication continued to transform world trade. In 1801 the United States and Scotland produced early steamships; by 1860 they left sailing vessels in their wake on the high seas. It took one year in 1650 to sail from Holland to Java; it took three months in 1850, and three weeks in 1920. World shipping increased fourfold from 1850 to 1910.

Britain had the world's first public railroad in 1830, but by 1845 the United States had twice as much track as Britain, and by 1914 the United States had half the world's railroad track. The first telegraph messages were sent between Baltimore and Washington in 1844. By 1866 a trans-Atlantic cable was laid down, and in 1870 a line from Britain to India reduced the time of conveying messages from eight months to five hours. By 1902 the British had cables worldwide. In 1860 telegraphs could send ten words a minute in Morse code; sixty years later they could send 400 words a minute. Electrification of the globe began around 1890.[16]

At the end of the nineteenth century another invention appeared to transform the world—the use of oil, a fossil fuel like coal laid down millions of years ago, to fuel internal combustion engines. A Scot, James Young, figured out how to refine crude oil in 1850, while Edwin Drake in Pennsylvania proved in 1859 that oil could be obtained by drilling through deep rock. Germans began to develop engines using oil in the 1880s. World production of oil rose from zero in 1800, to 20 million metric tons in 1900, to 3 billion metric tons in 1990, when people in the United States (4 percent of the world's population) were using 25 percent of the world's production. Oil would become a major strand, perhaps the major strand, of the story of the twentieth century.[17]

Imperialism and World Wars, 1850–1945

By 1870 Europe had about 70 percent of all world trade. By 1914 it occupied or controlled 80 percent of the world's area. In 1900 China had only 6 percent of the world's output, falling from 33 percent in 1800, and India had only 2 percent, falling from 25 percent in 1800. Africa had been parceled out to the European powers.[18]

In Europe and the United States the peak of racist thinking and policies based on racism marked this period. Racism exists ". . . when one ethnic or historical collectivity dominates, excludes or seeks to eliminate another on the basis of differences it believes are hereditary and unalterable," while at the same time professing to believe in human equality.[19] Racism seems to be mainly, if not exclusively, a product of Europe and the United States; its logic was fully worked out and implemented in three societies in the twentieth century: in the South of the United States against African Americans (1890s–1950s), by European colonists in South Africa against Africans (1910s–1980s), and in Hitler's Germany against Jews (1933–1945).

By the mid–nineteenth century many people of Europe and the United States regarded their dominance in the world as evidence of their innate biological superiority, rather than as indication of cultural, technological, or geographical advantage. France, Britain, Germany, Portugal, Belgium, and the United States used this racial ideology to justify acquiring new colonial territory.

The fact of their military power enabled the industrial powers to carve up much of the rest of the world among them in the decades before 1914. After the 1840s a significant imbalance existed in weaponry and communication systems; by the end of the 1800s the imbalance became even greater with the development of repeating rifles and machine guns and the medical ability to check diseases. The industrialized countries could take colonies with quick, cheap military action, and they chose to do so.

Britain took the lion's share of colonies; its empire spanned the globe by 1914, becoming the largest ever in the history of the world. (The Mongol empire was the largest continuous land empire.) India, its richest and most important colony, became Britain's gradually between 1750 and 1860, after the Moghul empire had lost vitality by 1710. Britain also took control of Canada, Australia, New Zealand, South Africa, Egypt, and enough else in Africa to control about 60 percent of its people by the end of the nineteenth century.

The African social order in the last decades of the nineteenth century was more firmly rooted in slavery than ever, which helped open the way to industrial powers, which took all parts of Africa except Liberia and Ethiopia (Fig. 12.2). The population of central Africa, which had had little prior contact with the outside world, fell perhaps a quarter between 1880 and 1920. Later in the century, with access to medical advances, Africa experienced the most rapid population growth the world has ever seen.[20]

China never became a colony; the Qing dynasty hung on until 1911 to 1912. But the situation became possibly worse than being a colony, beginning with Britain and the United States smuggling opium from India into China and leading to history's largest civil war, the Taiping Rebellion, from 1850 to 1864, in which 20 to 30 million Chinese died.

The United States fought a war with Spain in 1898 to take over Puerto Rico and the Philippines. Russia expanded into the Caucasus, while Japan took Formosa (Taiwan) and Korea from China, gained concessions in Manchuria, and took half the island of Sakhalin from Russia.

The ambitions of the imperial countries were aided by climate change in the late nineteenth century. Three times the monsoon rains failed for a period of three to six years, starting in 1876 to 1879. The lack of rain caused droughts and famines across the tropical countries and northern China, resulting in 30 to 50 million human deaths and contributing to the low industrialization of these regions. Africa has continued to be drought-prone in the twentieth century.[21]

The world system achieved by the Europeans, however, did not last long. It broke down in the twentieth century as the European powers fought among themselves, and the two non-Western industrial powers, Japan and Russia, competed with Europe and the United States for land and resources.

The end of European imperialism began with World War I, from 1914 to 1918, caused by the emergence of Germany as a great power competing with other European nations for colonies at a time of vigorous nationalism. An alliance of Britain, France, Russia, Serbia, and eventually the United States narrowly defeated Germany, Austria-Hungary, and the Ottoman Empire. The peace treaties slightly shrank Germany and imposed harsh

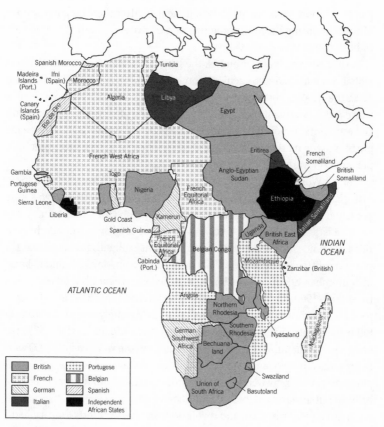

12.2   Africa in 1914

Legend:

British		Portugese	
French		Belgian	
German		Spanish	
Italian		Independent African States	

reparations on it for causing the war. In 1920 the victors established an international body, the League of Nations, with headquarters in Geneva, Switzerland, to manage and prevent future conflicts. Under its system of mandates, Germany's African colonies were given to various victorious nations. Parts of the Ottoman Empire were given to the victors, including Palestine to Britain. The rest of the Ottoman Empire collapsed in revolution in 1919 to 1923, and a secular Turkey emerged, which abolished the caliphate of Islam, leaving the Muslim world without a religious head or political center. The monarchy in Russia fell to revolution in 1917, and Italy's government collapsed in 1919 to 1927, giving rise to the dictatorship of Benito Mussolini (1883–1945). When the troops went home after World War I, an influenza spread around the world, killing about 40 million people, many more than had died during the war. After the vital participation of women during the war, some democracies began granting women the right to vote.

After this terrible war ended, the nations of the world retreated from global trade and tried to achieve some kind of economic self-sufficiency. With the crash of the U.S. stock market in 1929, banks that were linked to worldwide loans went bankrupt, causing worldwide economic depression. Governments responded by putting up tariffs and trying to minimize imports, making matters worse. By 1932 the world economy had shrunk by 20 percent and world trade by 25 percent.[22]

Since World War II ensued only twenty years after World War I, many historians see it as a continuation of the first war. Nationalism was rampant, and the ambitions of Mussolini in Italy, Adolf Hitler in Germany, and imperialists in Japan led them into aggressions that reignited the war. Britain, France, the USSR, and the United States allied again to defeat the Axis powers (Germany, Italy, and Japan) in Europe in May 1945 and in Japan in September 1945. Sixty million people died in the war, about 3 percent of the world's 1940 population. Six million were Jews in Germany and its occupied states, whom Hitler and his followers deliberately wiped out in mass killings. Twenty-five million were soldiers and civilians in the USSR, the country that sustained the highest casualties of the war. Over 100,000 were people killed in two Japanese cities—Hiroshima and Nagasaki—when U.S. pilots dropped a new weapon, atomic bombs made from nuclear energy, to bring about an early Japanese surrender.

The slaughter of wars and revolutions in the twentieth century had a profound effect on people's morale; before 1914 many had believed that the industrial world was too advanced to indulge in outmoded slaughter. Despite its horrors, the slaughter had little effect on the total population. Counting up wars, genocides, human-caused famines, and state terror campaigns, the total comes to about 180 to 190 million deaths, only 4 percent of total deaths in the century.[23]

Leadership by the United States, 1945–2000

By the end of the Second World War, the world's industrial powers had greatly damaged each other's industrial capacity, except that of the United States, which emerged to lead the peace process and to dominate the world's economy. For a brief period at the end of the war, the United States had a monopoly on nuclear weapons and half the world's industrial capacity. In the period following the war, remembering what happened to Germany in post World War I and responding to anxiety about the USSR and communism, the United States was able to underwrite the rebuilding of Europe's economy and to supervise the reconstruction of Japan. World industrial production boomed in the next half century.

The big breakthrough in energy in the twentieth century came as people learned to use oil as energy. The United States led the way in basing its economy on the use of oil, which revolutionized transportation, since cars and planes cannot run on coal. In 1912 Henry Ford developed the first electrified assembly line to produce automobiles; he had to pay his workers double to keep them at the monotonous job and, as a result, with two months' wages they were able to buy a Model-T Ford car. In the 1920s cars, telephones, and radios became widespread in the United States.

Under U.S. leadership, the world reglobalized again in the period from 1950 to 2000. The foundation for economic globalization was laid in 1944, when forty-five nations created the International Monetary Fund and the World Bank. Desperately wanting to avoid another devastating war, the victorious countries established the United Nations, with headquarters in New York City, to function where the League of Nations had left off, only more effectively. Under the leadership of Eleanor Roosevelt, in 1948 the General Assembly of the United Nations issued a Universal Declaration of Human Rights, the first formulation of the rights of all humanity, a milestone in human history. Two years later the United Nation Education, Science and Cultural Organization (UNESCO) issued a statement by leading scientists conceding that no scientific basis for the concept of race existed.

The competition for economic domination, however, did not end with World War II. The closest rival to the United States proved to be the USSR, which engaged with the United States in a cold war of economic and political rivalry, intensifying with the Soviet production of an atomic bomb in 1949 and the victory of communists in China in 1949.

The distrust between the USSR and the industrialized nations of the West dated back to the fall of the Russian monarchy in 1917 when, after a brief European-style government, Russians chose the Bolshevik Party to implement a communist revolution. The leaders of the USSR (fifteen republics) nationalized industry and agricultural production and provided housing, health, and education for their citizens. Using the ideas of Karl Marx, they believed that capitalist democracies were destined to collapse and become communist after violent class wars. (The word "capitalism" became common only in the twentieth century to distinguish it from "socialism" and "communism.")

The ideas of Marx found many adherents in Western democracies, as some people sought ways to diminish the inequality between workers and industrialists. In the USSR, however, the communist system did not keep up in the economic, military, and agricultural competition; by the late 1970s it could not feed its own population, and in the 1980s its finances were undermined by rapidly reduced oil prices. Party leaders desired the material benefits of capitalism. In 1991 the USSR disbanded as the other fourteen republics acquired their independence from Russia, whose leader, Michail Gorbachev, let them go peacefully.

The world wars of the twentieth century not only stopped European expansion but gradually broke up its former empires. After World War I the Austrian-Hungarian empire dissolved into four new nations: Hungary, Poland, Czechoslovakia, and Austria, while Ireland won independence from Britain. During and after World War II most of the world's colonies liberated themselves; there were about three times as many independent states in the 1990s as there had been sixty years earlier.

During the 1950s European countries, led by former rivals France and Germany, formed the European Economic Community to support each other and prevent domination of Europe by the United States and the USSR. This led, by the 1990s, to the semi-unification of Europe (after all these years!) as the European Union, with a common currency and common economic, agricultural, and migration policies.

Following World War II a serious problem arose in Palestine. During the First World War French and British diplomats had promised Palestine to both Jews and Arabs in exchange for support. After the war the League of Nations gave Palestine to Britain as a mandate. Britain allowed some Jewish migration to Palestine before World War II, and after it large numbers of Jews migrated there. Agitation for a Jewish state, plus American support, led Britain reluctantly to agree to the creation of the state of Israel in 1947 and 1948, as determined through the United Nations. Many Palestinians fled after some were killed; neighboring Arab states attacked Israel, which defeated them. Four wars took place between 1948 and 2005. Given the importance of Arab and Iranian oil to the United States and its allies, the U.S. has supported unpopular rulers of client Arab states, as well as providing Israel with a large portion of its military and financial support. Israel does not acknowledge developing its own atomic bombs, but is known to have done so, as this dangerous standoff continues.

The achievements of science in the last half of the twentieth century reached astonishing proportions. Antibiotics, developed during World War II, became available to save lives regularly. Scientists continued to find other life-saving medications; by 1987 the world's highest life expectancy was reached in Japan at seventy-eight years. Russians put the first artificial satellite (Sputnik) into orbit around Earth in 1957. U.S. astronauts landed on the moon in 1969, while in 1977 the Voyager I spacecraft was launched to travel past the outer limit of our solar system. In the 1950s U.S. scientists unraveled the genetic code in DNA molecules, nailing down Darwin's theory of evolution by showing the random mutation in genes. In the 1960s cosmologists found concrete evidence for their big bang theory that the universe began in a single explosive instant. Chemists produced plastics from oil residues in the 1940s. New strains of wheat, rice, and corn increased crop production two to four times from about 1960 to 1980.

Religions did not wither away as science grew in its power and prestige. Although secular outlooks increased in Europe and the United States, both Christianity and Islam expanded during the colonial period to lead the world in adherents by the end of the twentieth century. Religions emerged that emphasized the underlying unity of all religion, the similarity of all their messages—the Ramakrishna movement in India with its apostle Vivekananda (1863–1902) and the Baha'i faith, an offshoot of Persian Shia Islam, that adopted English as its preferred language. By the end of the twentieth century, in an atmosphere of fear and anxiety, fundamentalist versions of the world's religions experienced a revival. In 2002 there were estimated to be 10,000 distinct religions, with about 150 having at least 1 million adherents. If a group of ten people represented approximately the world's religions, three would be Christians, two Muslim, two unaffiliated or atheist, one Hindu, one Buddhist, and one representing all others.[24]

The last half of the twentieth century witnessed a spectacular sixfold growth of the world's economy. This seems normal to those who have lived through it, but such economic growth is historically unprecedented on a global scale and depended on science and technology as noted above, population growth, and increased energy use. The world population grew from 2.5 billion in 1950 to 6.1 billion in 2000. Oil production soared six times from 1950 to 1973. In the 1990s the average world citizen used energy equal to the power of twenty slaves, but that figure obscures the inequality in distribution of energy. The average U.S. citizen directed more than seventy-five "energy slaves," while a citizen of Bangladesh directed less than one. Nevertheless, between 1950 and 1975 inequality between the richest and the poorest regions narrowed, as well as inequality within industrial societies. The ancient gulf separating the rich and the poor in all previous urban societies actually decreased during this twenty-five year period.[25]

Since the 1970s, however, the inequality between the richest and poor-

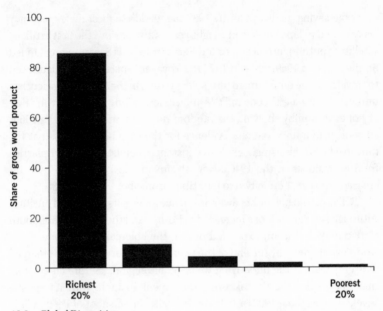

12.3   Global Disparities
(*Source:* Donella Meadows, Jorgen Randers, and Dennis Meadows, 2004, *Limits to Growth: The 30-Year Update,* White River Junction, VT: Chelsea Geen Publishing, 43.)

est areas, and between the richest and poorest people, has started to widen. After the 1980s the richest one tenth of people grew much richer, while the lowest one-tenth grew slightly poorer. By 2000 the income per capita of six nations containing almost half the world's population could hardly be plotted on a graph with that of the wealthiest nations. In 1985 the independent democracies, with a sixth the world's population, enjoyed five-sixths the world's wealth. In 2000 the richest 20 percent of the world's population controlled more than 80 percent of the world's gross product (Fig. 12.3).[26]

By the end of the twentieth century much of the world's wealth was no longer organized or regulated by national governments; it belonged to multinational corporations in some ways beyond the control of nations and wealthier than many nations of the world. No one knows what this development will mean for the future.[27]

Into this situation came the networked personal computer, possibly as significant an invention as the printing press. The electronic computer was first used in World War II to break codes; by the 1990s the networked personal computer emerged to widespread use. By 2000 there were several hundred million personal computers in the world, with 1.6 billion web pages to choose from, 78 percent of them in English. Computers put a premium on education; weakened the power of the state, at least temporarily; and increased communication among multinational corporations, academics, pressure groups, and terrorists, while increasing the possibilities for hackers to wreak havoc on the system.

More than 1 billion people in the world live without electricity, yet they are aware, from television in cafés or from radio, of what others have that they do not. The range of inequality in a world where rapid communication makes people familiar with their disadvantages creates an explosive situation whose unfolding cannot be foretold.

The fundamental facts, reduced to their simplest, are that during the twentieth century the human population grew almost fourfold, the world's economy grew fourteenfold, per capita income grew almost fourfold, while energy use expanded sixteenfold. This scale of expansion is something new under the sun, completely unprecedented in Earth's history.[28]

Unanswered Questions

1. What do "capital" and "capitalism" mean?

These are loaded words, whose meanings have been much debated. I have avoided using them as much as possible and can only suggest some of the possible nuances.

Capital originally meant money, but about 1770 in the ideas of the French economist Robert Jacques Turgot it took on the added meaning of the control of labor. In the nineteenth century Karl Marx used "capitalist" to mean those who control the means of production. Today people speak of fixed capital, or roads, bridges, canals, ships, tools, and machines, and variable or circulating capital, referring to raw materials, money, wages, and labor.

"Capitalism" came into widespread use only in the twentieth century in opposition to socialism and communism. One well-known historian of capitalism, Fernand Braudel, sees capitalism not simply as an economic system, or a free market, but as a social order supported by culture and the dominant classes, which are as important as government policy in sustaining capitalism. Long-term capitalism for Braudel includes the rise of towns and trade, the emergence of the labor market, increased population density, use of money, rise in output, and the international market.[29]

I see capitalism, socialism, and communism as variations of industrialization. Capitalism is industrialization by private initiative, while communism is industrialization by state initiative, with socialism a mix of both.

2. Is industrialization a good thing?

Historical accounts are often written assuming that industrialization is the goal toward which everyone is moving, the "march toward modernity," as Alan Smith puts it.[30]

Industrialization does seem to have been irresistible to people in areas where it could be achieved. With it come wealth, health, education, travel, stimulations, pleasures, and challenges of all varieties. A few, but very few, individuals and groups have refused these benefits when they could choose.

Industrialization, however, may prove to be unsustainable without colonies to exploit. Already industrial nations are having trouble keeping up their standard of living. With new nations trying to achieve industrialization, resources for doing so are becoming scarcer. Will industrialized nations be able to scale back gradually and/or find substitute resources? Perhaps the nonindustrialized nations will be in a better situation for confronting the challenges of the twenty-first century.

3. Did the culture of Protestant Christians play a crucial role in the initial creation of industrialization?

In 1904 the sociologist Max Weber argued, in an influential book, *The Protestant Ethic and the Spirit of Capitalism*, that Protestants held values and beliefs—hard work, saving, and rationality—that made them effective capitalists. For instance, Calvinists believed that making a fortune gave proof of God's election for salvation, which motivated people to work hard. Many Protestants believed that riches should not be used for conspicuous living, but for the public good. Weber based his idea on a survey, conducted in a section of Germany, which seemed to show that Protestants were wealthier and more engaged in economic activities than Catholics. This thesis seemed to fit the facts, namely the correlation between Protestant countries and areas where capitalists first flourished (Holland and Britain) and the tendency of the Catholic Church to uphold traditional societies.

In the century since its publication, Weber's classic has been debated so thoroughly in academia that the debate has been called "the academic Hundred Years' War."[31] By the late twentieth century Weber's thesis seemed refuted by further developments. Russia and Japan, with their varying religious beliefs, achieved highly industrialized societies early, and after mid-century several Asian "tigers" did—South Korea, Singapore, Taiwan. Some nations, however, have industrialized more slowly, if at all; theorists debate whether the reasons might be geographical, structural, or cultural, and to what degree. Weber may have correctly identified the original motive for working harder; later transitions to industrialization may be motivated by the rewards of industrialization itself, which could not yet be seen in the first experiment in Europe.

13

What Now? What Next?

Historians do not usually attempt to describe the present; they leave that task to sociologists, political scientists, and politicians. Yet I have not been limited up to now by what historians usually do. Analyzing the present to plan for the future is part of human capability and human responsibility, so here goes.

Some Global Measures

In the year 2000 there were 6.1 billion people living on Earth—6 to 12 percent of all the estimated 50 to 100 billion who have ever lived. By many measurable indicators the lot of people now alive has much improved by comparison with earlier times.[1]

In 1900 the global average lifespan hovered about thirty years, not much different from the twenty-two-year lifespan of an average citizen in imperial Rome. By 2000 the global average lifespan had reached sixty-seven years, with extra years of good health rather than suffering.

Food prices declined notably in the last half of the twentieth century; in 2000 food in general cost to consumers less than a third of its price in 1957 (Fig. 13.1). This happened primarily because of high-yield crops, irrigation and dams, fertilizers and pesticides, and the management skills of farmers, not counting the cost in environmental damage. The increase in cheap calories was not shared by sub-Saharan Africa, which increased its calories per day by only about 150 from 1960 to 1997, compared to Asia's increase of 800 calories per capita per day. Sub-Saharan Africa used much less

13.1   The Cost of Food, 1957–2000
(*Source:* Bjorn Lomborg, 2001, *The Skeptical Environmentalist: Measuring the Real State of the World,* London and New York: Cambridge University Press, 62. Includes material originally published in the International Monetary Fund, *International Statistics Yearbook,* 2000.)

fertilizer per hectare than Asia and irrigated only 5 percent of its cultivated area, compared to 37 percent in Asia, with more soil erosion. Sub-Saharan farmers have potential for producing more food, but have been hampered by severe ethnic conflicts, droughts, corruption, inadequate infrastructure, poor education, and fixed farm prices.

During the twentieth century people on the average became much wealthier than in all previous history. Global production per capita, i.e., global gross domestic product per capita, remained at about $400 until the 1800s, when it doubled to $800. By 2000 it exceeded $6,000. While this average includes the world's wealthiest members, it is also true that the proportion of poor people is decreasing somewhat, especially in southern and eastern Asia. Estimates of the proportion of the world's poor in 1950 suggest about 50 percent. World Bank figures in 2000 showed a decline of the poor in the Third World from 28.3 percent in 1987 to 24 percent in 1998. Since 1950, some 3.4 billion people may have risen above extreme poverty.

By the end of the twentieth century people engaged in more formal education than at the beginning. In the developing countries the average number of years in school increased from 2.2 in 1960 to 4.2 in 1990, while the average in the Western world increased from 7 in 1960 to 9.5 years in 1990. During the twentieth-century India stands out in educational gains,

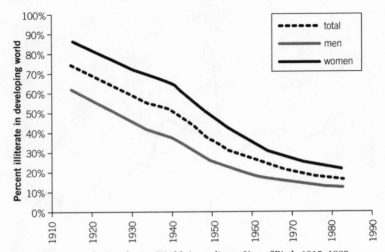

13.2   Illiteracy in the Developing World, According to Year of Birth, 1915–1982
(*Source:* Bjorn Lomborg, 2001, *The Skeptical Environmentalist: Measuring the Real State of the World,* London and New York: Cambridge University Press, 81. Includes material originally published in the *Compendium of Statistics on Illiteracy–1990 Edition,* Paris: UNESCO, Office of Statistics.)

with secondary school enrollment rising from 3 to 4 percent in 1900 to 50 percent in 1998 and literacy rising from 20 percent in 1900 to 65 percent in 2001. Nonliteracy in the whole developing world, measured from date of birth, sank from 75 percent in 1910 to about 17 percent among the youth of 2000 by UNESCO figures (Fig. 13.2). The overall rate of nonliteracy of people of all ages in the developing world in 2000 was about 30 percent.

The rapid rise in the annual rate of growth of the human population peaked in 1964 at just over 2.17 percent and fell by 2000 to 1.26 percent, with a doubling time of fifty to fifty-five years. A birth control pill became available in 1960, and the Chinese government mandated in 1979 a one-child quota per family. The average number of children per family in the developing world was 3.1 in 2000, down from 6.16 in 1950, correlated with birth control and women's education. According to UN estimates in 2000, using the medium variant in fertility, we can expect almost 8 billion people by 2025, about 9.3 billion by 2050, and a stabilized population at about 11 billion in 2200, with most of the increase in the less developed regions (Fig. 13.3).

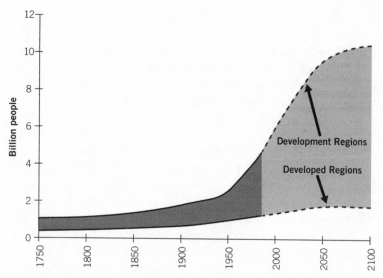

13.3 Projected World Population
(*Source:* Paul Kennedy, 1991, *Preparing for the Twenty-First Century,* New York: Ballantine, 23. Includes material originally published in *The Economist,* January 20, 1990, 19.)

Experiment with Earth

As humans used their ingenuity over the centuries to increase their numbers, their life span, and their income, they inadvertently began to conduct an experiment with the planet that sustains them. In 1972 a group called the Club of Rome, composed of scientists, educators, industrialists, economists, and civil servants from ten nations, issued a warning called *The Limits of Growth.* By the end of the twentieth century many informed people had developed a deep concern about the predicament of humanity on a limited planet.[2]

For example, the human population in the last century grew from 1.6 to 6.1 billion. Even if people limited themselves at once to replacement-size families (not a realistic expectation), population growth would not stabilize before reaching at least 8 to 9 billion people. Most of this increase will occur in nations not yet industrialized and least able to support it. This urgently raises the question: When will the limits of the Earth's carrying capacity be reached or have they been already?

The human experiment with its planet has many dimensions, all interconnected in the living organism of Earth. For brevity and clarity, I am sorting these into the categories of air, forests, soil, water, and radiation.[3]

Air

By 2000 half the world's people lived in cities; Britain passed this mark in 1850, China will pass it sometime between 2005 and 2010. Tokyo, Japan, stood as the world's largest city, with 34 million people, as many as lived in the whole world when agriculture began.

When the World Health Organization looked, in 1988, at the quality of air in cities, it estimated that 1 billion of 1.8 billion city dwellers breathed unhealthy levels of sulfur dioxide, soot, and dust. Cities like London and Pittsburgh, in countries that industrialized early, were wealthy enough to clean up their air significantly, but the mega-cities (over 10 million) of the late twentieth century grew too fast, could not enforce their regulations, and gave precedence to economic development. Such cities include: Mexico City (18 million), Calcutta (15 million), Shanghai (13 million), Beijing (11 million), Karachi (10 million), Cairo (12 million), and Seoul (10 million). In these cities atmospheric pollution is sufficient to kill several million people annually through respiratory disease. In Los Angeles, too, smog remained in the 1990s a regular health hazard, the most serious urban air pollution problem in the United States.

Another human addition to the atmosphere has been chlorofluorocarbons, or CFCs. The first CFC, freon, replaced inflammable and toxic gases in refrigeration and made air conditioning possible. But producers of freon did not think about what happens when CFCs reach the upper atmosphere, namely, that ultraviolet (UV) radiation breaks up their molecules. This releases agents that break up the molecules of the thin layer of ozone (created by the interaction of oxygen and sunlight) that protects life on Earth from the dangers of UV radiation. In 1974 scientists suggested the theoretical possibility of this occurring; observations in 1985 confirmed that it had indeed occurred over Antarctica. After 1988 the worldwide use of CFCs declined 80 percent; nations acted swiftly in the Montreal Protocol (1987) to ban CFCs because the effects of increased UV radiation would be devastating: it kills plankton, the basis of ocean food chains, affects photosynthesis, and in humans, causes cataracts, skin cancer, and suppressed immune systems. However, the CFCs already released will continue to destroy ozone for at least a decade or two into the twenty-first century, and the ozone layer will take decades to build up again. No one knows what the long-term effect on immune response and human health will be.[4]

Other damaging human additions to the atmosphere are the so-called "greenhouse gases"—chiefly carbon dioxide and methane, a gas produced by decomposing vegetable matter and contained in natural gas. Our climate is controlled by the composition of the atmosphere, and these green-house gases keep the Earth warm by trapping the solar rays reflected off the surface of the Earth.

Before 1800 the level of carbon dioxide varied from about 270 to 290 parts per million. Humans contribute to this level by burning fossil fuels (coal, oil, and natural gas) and by deforestation. After 1800 the level of carbon dioxide began to rise noticeably, and by 1995 had reached about 360 parts per million. Fossil fuel consumption caused three-quarters of the increase, deforestation one-quarter. In 1990 the U.S. accounted for 36 percent of the industrial nations' carbon dioxide emissions, followed by a rise in its greenhouse gas emissions of 13.1 percent between 1990 and 2002. Methane has increased since 1800 from 700 to about 1,720 parts per billion; the main sources were livestock (which emit methane from their digestive tracts), garbage decomposition, coal mining, and fossil fuel use. The melting of the Siberian tundra may also be releasing enormous quantities of methane (Fig. 13.4).

The Earth warmed up modestly in the twentieth century; the average surface temperature increased by 0.3 to 0.6 degrees Celsius, while the temperature of the oceans increased much more. The greatest warming took place north of 40 degrees latitude, north of Philadelphia, Madrid, and Beijing. By the end of the century most scientists agreed that climate change

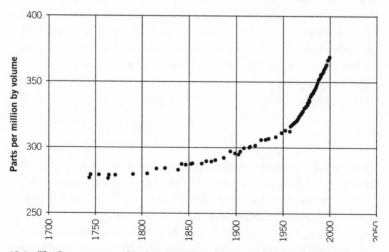

13.4   The Concentration of Carbon Dioxide in the Atmosphere
(*Source:* Donella Meadows, Jorgen Randers, and Dennis Meadows, 2004, *Limits to Growth: The 30-Year Update,* White River Junction, VT: Chelsea Green Publishing, 7.)

was occurring and that human activities were contributing to it. Most expected that temperatures would rise in the twenty-first century, anywhere from 2 to 10 degrees Fahrenheit (1 to 5 degrees Celsius), with unknown but sharp consequences, among them more droughts and floods, rising seas, expanded tropical diseases, accelerating extinctions, and/or a slowing of the Gulf Stream in the North Atlantic. By 2006 leading scientists calculated that people must begin to reduce their carbon emissions significantly within the next decade, or face unavoidable climate change with far-ranging undesirable consequences.[5]

Forests

The major impact on forests came from their use as fuel. By the end of the eighteenth century Britain faced serious levels of deforestation; its levels of forest cover (5 to 10 percent) and per capita supplies of wood were lower than in densely populated China and India. The forest level of Britain and western Europe stabilized between 1800 and 1850, rescued by forest reserves and new crops from the Americas, better forest management, and by the use of coal.[6]

Estimates of the reduction of global forests since 10,000 years ago vary from 15 to 50 percent. In Africa and monsoon Asia probably only one-third of the land covered in forests 10,000 years ago remains in forest. In Russia two-thirds remain, in the Americas three-fourths. One half of the total clearance occurred in the twentieth century, and one half of that in the tropical belt since 1960. Only three large blocks of forest remain on Earth—across Northern Eurasia from Sweden to Sakhalin, across Northern America from Alaska to Labrador, and the Amazon and Orinoco basins of South America. At the present rate of conversion one fourth of forests that remain will be converted to other uses in the next fifty years.[7]

Soil

Earth's crust is covered by a thin layer of topsoil, up to hip deep, that takes centuries or millennia to build up. This top layer easily erodes after deforestation, which occurred rapidly in the twentieth century as the world's area under cultivation roughly doubled, much of it in the rainforest regions. Soil erosion in Africa averaged eight to nine times greater than in Europe; only in Africa did food production decline after the 1960s. Elsewhere erosion and loss of soil was masked by higher food production, forced through the use of chemical fertilizers to replace the depletion of nitrogen and phosphorus from the soil that limits crop production.

The advent of manufactured fertilizers plays a significant role in the recent human story. Manufactured fertilizers are created by extracting nitrogen from the air by synthesizing ammonia. An academic chemist, Fritz Haber, in what is now Polish Silesia, figured this out in 1909. An industrial chemist, Karl Bosch, worked out the mass production of fertilizer, now called the Haber-Bosch process. It depends, however, on burning large amounts of oil. Hence, the great leap in food production, which allows an extra 2 billion of us to eat, comes at high cost: dependence on oil and pollution of soil and water caused by the artificial fertilizers, which have increased in use from 4 million tons in 1940 to nearly 150 million tons in 1990. The use of fertilizer and new strains of plants increased per capita grain production from 1960 to 1980, but it has been declining slightly since.

Since the production of food depends on fertilizers, which depend on burning oil, the cost of food reflects the cost of oil. Known oil reserves are estimated to last for about forty years at the current rate of use. The United States in 2005 consumed one out of four barrels of oil produced worldwide. Global demand is rising rapidly as China and India industrialize. No one knows when the peak production of oil will be reached, after which demand will force prices up.

Water

The health and wealth of any society depend on having adequate supplies of clean water. People used a great deal more water at the end of the twentieth century than ever before. By 1920 the richer cities of the world were able to provide safe drinking water to their citizens, but this occurred unevenly or not at all in Asia, Africa, and Latin America. Colonial cities often acquired water filtration and sewerage systems in the European quarters only, not throughout the city, as happened in Shanghai, Kampala, and Algiers. In 1980, half the world's urban population had no wastewater treatment of any kind, and many urban residents had no sewerage hookup.

Rivers and inland seas in industrial areas acquired toxic loads of biological and chemical wastes in the twentieth century. The Ganges River in India carried by 1990 the waste of 70 million people, plus several tons of cremated human remains annually, roughly 60,000 animal carcasses, plus industrial emissions and phosphorous from fertilizer. Government cleanup efforts have had little effect. On the other hand, cleanup of the Rhine River in Germany since World War II has proved effective enough to restore salmon to the river. After 1975 all the countries around the Mediterranean Sea, except Albania, convened under the United Nations Environmental Programme (UNEP) to manage their wastes going into the sea, in an action called the Mediterranean Action Plan (MAP). Twenty years later the sea is more polluted than before, but less than it would have been without MAP.

The oceans have seemed, by their vastness, to be immune from pollution, but no longer. By 1992 plastics accounted for 60 percent of beach litter; reports are coming in that they now litter the ocean floor and are being ground by water action into the molecules of the sea and its inhabitants. Mercury levels of larger fish now make them unsafe for human consumption. Fish used to feed the poor, now they feed the rich. For instance, the number of blue fin tuna declined 94 percent from 1970 to 1990 and cost $100 per pound in Tokyo in the early 1990s. Cod, long the staple of Europe, sold for $80 a pound in Stockholm in 2002. One third of fish consumed in 2002 were farmed on coastal wetlands, but this destroys wetlands, puts food wastes and antibiotics into the water, spreads viruses, and permits cultivated species to escape into the wild.[8]

The increased use of water and the increased area that has been paved over have lowered the water table in many areas of the world. In California's central valley, groundwater is depleted by an average of one cubic kilometer a year. The underground water supplying irrigation in the midwestern United States is being depleted at 12 cubic kilometers a year. In northern Africa and the Middle East, water is being pumped from desert aquifers that receive no new supplies. In India's agricultural states, the water table is subsiding half a meter a year. In northern China the Yellow River is drying up, partly due to overpumping of wells. It seems fair to conclude that limits posed by polluted and used-up water will likely constrain human activities in the twenty-first century.[9]

Radiation

During World War II U.S. scientists created a pollutant problem never seen before on Earth—radiation from splitting atoms of uranium or plutonium. When World War II ended, only the United States had developed the atomic bomb; four years later the USSR had done so. By 2005 seven countries admitted having nuclear weapons (United States, UK, France, Russia, China, India, and Pakistan), and two countries (Israel and South Africa) had them without admitting it. South Africa apparently has dismantled its nuclear weapons. Several dozen other countries by 2004 had the capability to make weapons-grade fuel, since the fuel rods of uranium or plutonium that are used to produce electricity can also be used to produce the enriched uranium or plutonium needed for bombs.[10]

The only atomic bombs ever dropped on human populations occurred when the U.S. dropped two on Japan in 1945 to end World War II, without knowing what the effects of radiation would be, using Japanese civilians as guinea pigs. Radiation causes acute illness and death, depending on the exposure. It also causes long-term consequences if one survives initial exposure: leukemia and other cancers and increased genetic mutations.

Since 1945 the U.S. has built tens of thousands of nuclear weapons and tested more than 1,000 of them in vacant territory. Its main bomb factory was the Hanford Engineering Works on the Columbia River in south central Washington, which built the bomb that destroyed Nagasaki. Over the next fifty years, Hanford released billions of gallons of radioactive wastes into the Columbia River and leaked some into the groundwater. A bomb test at Hanford in 1949 released radiation at levels between 80 and 1,000 times the limit then thought tolerable. Local citizens did not learn of these tests until 1986. Partial cleanup of fifty years of weapon production in the United States is projected to take 75 years and to cost $100 billion to $1 trillion, the largest environmental project in history; full cleanup is impossible.

The Soviets built a large nuclear weapon complex in a few years, doing most of their testing in Kazakhstan and on the Arctic island of Novaya Zemlya. They dumped much of the nuclear waste at sea, mostly in the Artic Ocean. Their center for reprocessing used nuclear fuel, the Mayak complex in the upper Ob River basin in western Siberia, is now the most radioactive place on Earth. It accumulated fifty times more plutonium than the Hanford site—26 metric tons. Liquid wastes, stored at Lake Karachay, were exposed during a drought in 1967, when winds sprinkled dust with 3,000 times the radioactivity released at Hiroshima over half a million unsuspecting people. Radiation remains in the air, soil, and water.

Attempts have made, starting in the USSR, Britain, and the U.S., to use nuclear power to generate electricity. By 1998, twenty-nine countries had some 437 nuclear power plants in operation, but not one made commercial sense, since they all survived on massive subsidies. One kilowatt hour of nuclear energy cost 11 to 13 cents in 1999, compared to 6.23 cents for fossil fuel energy. The used fuel rods continue to be radioactive, and no safe place to store them has been found.[11]

Closing down nuclear plants proved expensive. Mishaps occurred, climaxing at Chernobyl in Soviet Ukraine in 1986, when human error led to an

electrical fire and explosion that nearly destroyed one reactor. The total release of radiation was hundreds of times greater than that given off by the
bombs in Japan. Three quarters of a million soldiers were involved in the
cleanup, all exposed to cancer-causing radiation; about 135,000 people had
to leave their homes indefinitely. The countries most affected have been
Ukraine, Belarus, and Russia. Severe contamination of food occurred;
blackberries sold in Moscow markets in 2003 still tested radioactive. Everyone in the northern hemisphere received some radiation from Chernobyl;
part of its fallout will be lethal for 24,000 years.

At the beginning of the twenty-first century technological hopes are
focused on developing a way to fuse two atoms of hydrogen into a single
atom of helium and to use the extra energy. The fuel would be seawater,
with some radioactive waste or emissions. Despite investments of $20 bil-
lion, no way has yet been found to exceed more than 10 percent of the astronomical temperatures required for producing fusion energy.[12]

The combination of humanity's effects on Earth's environment is causing a rapid extinction of many current species. Rapid extinctions have happened before. The geologic record reveals at least five, the most extensive
ones 250 and 65 million years ago; only 1 to 10 percent of all the species that
have ever lived are alive today. Many scientists believe that humans are
now perpetrating a sixth major extinction.[13]

From the 1960s on, there have been thousands of international agree-
ments about environmental problems, having considerable impact. Problems that were politically and technologically the easiest to address have
been ameliorated: industrial wastewater, sulfur dioxide emissions, leaded
gas, sewerage treatment. Other problems, however, have intensified. Toxic
farm runoff and nitrous oxides from vehicle exhausts have increased, and
industries have fought back.

In the 1980s some poorer countries, especially Brazil, Kenya, and India,
established environmental programs. The Green Belt movement in Kenya,
led by Wangari Maathai (b. 1940), has planted 30 million trees in twenty-
eight years. In 2004 the Nobel Committee in Sweden awarded her the year's
Nobel Peace Prize in recognition that only with sufficient resources can
peace prevail.

In 1992 the United Nations convened in Rio de Janeiro, Brazil, a Conference on Environment and Development. It reached the first international
agreement that growth must be balanced by sustainability, but in practice
it made little headway. The United States stood firm that its lifestyle was
not up for negotiation. Brazil insisted on developing its Amazon rainforest;
China and India did not give up their industrial ambitions. A follow-up

conference in Johannesburg in 2002 produced even less; participants were almost paralyzed by competing interests. Pollution, deforestation, and climate change, since gradual, do not seem as immediately threatening to people and their leaders as do the lack of economic growth or the lack of military preparedness. In the midst of enormous global disparities people in wealthy countries do not seem to believe that their security is threatened by the despair of people in poor countries.

Possible Short-Term Scenarios

We humans are in a serious predicament, but how dire is it? No one knows. It may already be too late to prevent a calamitous fall-off in population and urban life. It may be that we can still prevent a fall-off, but only if we act decisively within the next twenty years. Or it may be that economic development and new discoveries and technologies will enable us to change gradually to a sustainable global society, using our present policies.

Three of the system analysts who published *The Limits to Growth* in 1972—Donella Meadows, Jorgen Randers, and Dennis Meadows—continued to model data on computers, changing one variable at a time and watching the likely effects on other variables. After thirty years of refining the data and their analysis, they published a second report, *Limits to Growth: The 30-Year Update,* based on information available through 2002. They found that the actual numbers of population and food production coming in over the past thirty years have matched quite closely their earlier scenario of what would happen if current policies continued, with population growth decreasing somewhat more than expected. They reach the conclusion that what they call the "human footprint" outstripped the carrying capacity of Earth in the 1980s. Their evidence includes the following: grain production appears to have peaked in the mid-1980s, there are no prospects for increase in the harvest of fish; costs of natural disasters are rising; conflicts have developed in efforts to allocate fresh water and fossil fuels; the United States and others continue to increase greenhouse gas emissions; there is economic decline in many areas and regions (fifty-four nations with 12 percent of the world's population).[14]

Many voices in the U.S. disagree with, or deny, the analysis given in *Limits to Growth.* These dominant voices say that more growth is the answer to our problems, that the operation of the free market and resulting advances in technology and innovation will surmount the challenges facing humanity.

For example, a professor of business administration at the University of Maryland, Julian Simon, believed that the twentieth century was the start of a long-term trend of improved life on Earth. Simon (d. 1998) served as an adjunct scholar at the Cato Institute in Washington, D.C., a public policy research foundation dedicated to liberty, peace, and limited government. The Cato Institute published posthumously Simon's book, *It's Getting Better All the Time: 100 Greatest Trends of the Past 100 Years* (with Stephen Moore), in which the authors predicted:

> 1. The wealth and health that have been attained in the U.S. today will be spread to the rest of the world in the next 50 years. We are in the first stages of a worldwide boom in wealth. 2. The price of natural resources will continue to fall, representing less constraint on growth than ever before. 3. Continued improvements in agriculture—particularly in the area of bioengineering—will mean bountiful food production that far outpaces population growth.[15]

In rebuttal to this kind of optimism, Meadows, Randers, and Meadows say that the optimists do not count the cost to the environment of human achievements. The people who believe there are limits to growth cannot be positive yet that their projections are realistic, even though the past thirty years seem to confirm them. They expect that the next ten to twenty years will produce clear evidence as to whether their projections, or those of the optimists, are more nearly accurate; they promise a forty-year update in 2012.

What kind of collapse could occur if rapid growth and pollution continue, if overshoot occurs? According to the scenario generated in *Limits to Growth: The 30–Year Update,* the world's industrial output might peak about 2015 to 2020, then fall back to the level of 1900 by 2100. World population might peak about 2025, then fall steeply back to somewhat above the level of 1900 (1.6 billion) by 2100. Other variables (life expectancy, food per person, consumer goods and services) are forecast to return approximately to 1900 levels by 2100, except resources, which slide to about one-quarter of their 1900 level. How this would actually play out in the real world is anybody's guess.

If, instead of continuing present trends, the human community mobilized sufficiently to cooperate in unprecedented ways to meet the unprecedented threat, what might be possible? The scenarios generated by Meadows, et al, suggest that a sustainable world society could still be achieved that would have a population of about 8 billion living at standards equivalent to those of the lower-income nations of present day Europe.

In order to achieve this sustainable society, people would need to act immediately in three dimensions simultaneously, namely, to limit population, to limit industrial growth, and to improve technologies. To limit population, every couple would need access to birth control methods and would limit their children to an average of two per family, as roughly 1 billion people in the most industrialized countries have already done. To limit global industrial growth, output per capita would need to stabilize at about 10 percent higher than the world average in 2000, equitably distributed. This would be a leap ahead for the world's poor and a significant adjustment for the wealthy. (If income is not distributed more equitably, then it seems unlikely that population growth and migrations can be stabilized.) Finally, technologies would need to be developed and implemented to increase the efficiency of using resources and to decrease pollution and soil erosion.

Meadows, Randers, and Meadows estimate that if these measures had been implemented twenty years ago, the population could have been stabilized at 6 billion, with more resources for everyone. If people wait another twenty years to implement these policies, these analysts believe it will be too late to avoid steep declines.

What are the prospects for the international cooperation that would be needed to achieve a sustainable world society? The current network of international cooperation and communication is certainly at historically unprecedented levels. One case study in international cooperation stands out, the ban on CFCs described briefly earlier. This case bears re-examination because for once consumers acted promptly, some governments and corporations acted as courageous leaders, and the United Nations provided sophisticated negotiation to solve an international problem that threatened life on Earth.

The process of cooperation involved a complex array of participants. As soon as the problem became known, environmental groups in the United States campaigned against aerosol cans, and by 1975 their sale dropped by 60 percent. Two months after the first UNEP conference in Vienna in 1985, scientists first documented the ozone hole over Antarctica, caused by circumpolar winds that keep the air in place for months before it disperses around the globe. The U.S. took leadership, despite deep divisions in President Reagan's administration. At the next conference, in Montreal in 1987, the major CFC producers agreed to freeze production, then to cut back 30 percent by 1998. Scientists produced more evidence of ozone depletion; DuPont Corporation, the largest U.S. producer of CFCs, agreed in 1988 to suspend all production. Substitutes for CFCs in refrigerators and spray cans, relatively cheap to implement, were found. The director of UNEP, Mustafa Tolba, proved a highly skilled negotiator. Continued measurements showed ozone depletion occurring nearly twice as fast as expected. By 1992 in Copenhagen the signers of the Montreal Protocol agreed to end all CFC production by 1996, and by 1996 157 nations had signed the agreement. The final cost, including negotiation and enforcement, came to about $40 million. By 2002 the Scientific Assessment of the Ozone Layer, done by UNEP and the World Meteorological Organization, reported that ozone levels should be increasing by 2010 and return to pre-1980 levels by mid-century. The peak of ozone depletion, if this scenario proves correct, will have been from 1995 to 2010. Three countries (Russia, China, and India) are permitted to produce some CFCs until 2010, and some unknown amount of smuggling is still going on. This seems to be a success story of international cooperation stopping overshoot and easing the planet back to sustainability of its ozone layer.[16]

On the other hand, in 1997 international representatives negotiated the Kyoto Accords, an agreement to reduce carbon dioxide production by 5 percent of its 1990 levels by 2008. The U.S. negotiators agreed to a 7 percent reduction in U.S. emissions, but the Senate refused to ratify the treaty by a vote of 95–0. Enough other nations signed for the treaty to take effect in 2005. Without the United States, the reduction is likely to be about 1 percent. Will Europe and Japan be able to reduce emissions without hurting their economies? Within the United States, will states like California take the initiative to reduce emissions, led by industries that recognize climate change as a significant threat to their profits?

People have a chance of mobilizing for a sustainable future only if they become aware of the dangers of not changing course. As early as 1972, U Thant, the Burmese secretary-general of the United Nations, tried to inform government officials and the general public:

> I do not wish to seem overdramatic, but I can only conclude from the information available to me as Secretary-General, that the Members of the United Nations have perhaps ten years left in which to subordinate their ancient quarrels and launch a global partnership to curb the arms race, to improve the human environment, to defuse the population explosion, and to supply the required momentum to development efforts. If such a global partnership is not forged within the next decade, then I very much fear that the problems I have mentioned will have reached such staggering proportions that they will be beyond our capacity to control.[17]

Twenty years later, in 1992, a group of 1,600 scientists from seventy countries issued their warning, calling it "World Scientists' Warning to Humanity":

> Human beings and the natural world are on a collision course. Human activities inflict harsh and often irreversible damage on the environment and on critical resources. If not checked, many of our current practices put at serious risk the future that we wish for human society and the plant and animal kingdoms, and may so alter the living world that it will be unable to sustain life in the manner that we know. Fundamental changes are urgent if we are to avoid the collision our present course will bring about.[18]

What are the chances that humans can find ways to manage and guide their experiment with Earth, now either a bit or a lot out of control? No one knows, but there seem to be three basic possible resolutions of this experiment, resolutions that will likely occur within the lifetimes of present children and young adults. Either people will be able to curb their growth and use of resources, or nature and human nature will do it for them (disease, starvation, warfare, genocide, and social collapse), or some combination of both. The three authors of *Limits to Growth* hold different opinions about which scenario is most likely; each took one of these three possible positions.

Humans have frequently been in difficult predicaments, yet they have often risen to the challenge. Massive migrations and diebacks in population have occurred repeatedly in human history, requiring human ingenuity to find fresh ways to adapt and persevere.

Our current predicament, however, seems new in human history. It comes, especially for those in industrialized countries, after 500 years of rising living standards, which have supported us in believing that we can count on human development and progress. It differs from earlier human predicaments in that we no longer have areas uninhabited by people into which to move. It involves likely changes in climate that will impact the whole planet, not just local parts of it. We have inserted ourselves into the very process of evolution as agents, knowing little of what we are doing, following our instincts for survival.

When we look at the big picture, we see that our capacities and our social behavior have developed over at least several hundred thousand years, evolving as we lived as hunter-gatherers in small groups. Our evolution was not exactly speedy, in terms of our lifetimes. Our complex cultures create significant inertia; we tend to hold on to our patterns of living to the extent that we can. Culturally we evolve, but usually not rapidly. It took thousands of years to work out the transition to farming and to living in cities, which we are still working on. We have been industrializing for a mere 300 years.

Can we evolve culturally fast enough to make the transition to sustainability? Can we find a way to avoid a precipitous crash in our population? Can we make peace with Earth before it forces us into submission? If we wait until the data are unambiguously clear, our choices seem likely to be seriously compromised. What can propel humans to act before we are confronted with massive, immediate danger?

The Universe Abides

As a species, we humans are young. If we have a potential future of, say, 1 million years, then we are adolescents, ten to twenty years old as a species. Is there some way to think about what may happen to us during the next several hundred to several thousand years, the middle-range future?

Making predictions this far out becomes absurd, yet people speculate and dream. Scientists work to find new sources of energy, including possibly hydrogen fusion. Biotechnologists may find new ways to feed and clothe 10 to 12 billion people. Genetic technologists may soon start to manipulate human genetic makeup, without waiting for the slow process of natural selection.

Some scientists believe we can domesticate the moon and nearby planets and move some people there, or have colonies of people living in some kind of space arks. A probe landed on Mars in 1997 and carried out experiments. Several more probe missions are planned for the first decade of the twenty-first century. An American engineer, Robert Zubrin, worked out a plan in the early 1990s for a manned mission to Mars (four astronauts, thirty months) estimated at that time at a cost of $30 to $50 billion. The U.S. Senate, however, in 1969 put an end to government funding for manned space missions, after reviewing the costs of the Apollo mission to the moon—$25 to $30 billion at the time, worth more than a $100 billion in 1998.[19]

The middle-range future is best explored in works of fiction. George Stuart describes in *Earth Abides* (1949) a world in which a disease sweeps away almost everyone, with a small pocket of humans starting afresh, without contact with any other humans. Walter Miller, writing at the height of the Cold War, portrayed a future of periodic nuclear holocausts in *A Canticle for Leibowitz* (1959). A trilogy by Kim S. Robinson explores what colonizing Mars might be like: *Red Mars* (1991), *Green Mars* (1994), *Blue Mars* (1996).

Beyond the middle-range future lies the long-range future, which changes so slowly that astronomers are willing to hazard predictions about what is in store for planets, the sun, and the universe. They know that our sun is almost halfway through its cycle; having burned for about 5 billion years, it has about 5 billion years to go before its long death struggle begins. As the sun ages, its luminosity is increasing 10 percent every billion years, thus slowly heating up the surface of Earth. In some 3 billion years Earth will be receiving as much heat as Venus does now and will have long since become too hot to sustain life.[20]

When the sun burns all its hydrogen in 5 billion years, it will become unstable, ejecting material from its outer layers. Its core will expand until it reaches out into the space in which Earth now revolves. With less gravity from the sun, Earth will drift into an orbit about 60 million kilometers farther out. When the sun begins to burn its helium, it will shrink again, then flare up again as it manufactures oxygen and carbon, then shrink at last into a white dwarf as its furnaces die down, cool, and darken.

Our nearest large galaxy, Andromeda, is currently approaching the Milky Way and will meet it about when the sun begins to swell up. Andromeda will move away again and return ever closer until, after a few hundred million years, Andromeda, the Milky Way, and other small local galaxies will merge into one system.

Already the era of star formation in the Milky Way is drawing to a close. About 90 percent of the material from which stars are formed has already been used in our galaxy. In a few tens of millions of years stars will be dying, and the lights will be going out in our universe. There will not be sufficient energy to create new complex entities. In a few hundred billion years the universe will become a graveyard of dark, cold objects—brown dwarfs, black holes, dead planets.

As the universe expands it will become, some 10^{30} years from the big bang, a cold dark space with black holes and stray subatomic particles dancing light years away from each other. Even black holes may eventually disappear. The only structures left in the endless night of the universe will be subatomic particles: electrons, positrons, neutrinos, and photons.

(The above scenario is based on present knowledge, which points to an endless open universe. If, however, enough dark matter is located with the potential to stop the expansion of the universe, then a closed universe could occur, which would reverse its expansion to close back on itself in a reverse repeat of its history, back to conditions similar to the big bang.)

Since we can now imagine with some accuracy the birth and life of our whole universe, the 13 billion years covered in this book no longer seem an eternity. They represent the burst of creativity at the beginning of our universe, its childhood years, when its immense energy and warmth produced the glorious entities we experience. We ourselves, still in the childhood of our potential, are the result of the universe's youthful creativity and power.

But the youthful time of our universe will come no more. In its exuberance our universe has produced a creature of such extraordinary complexity that we seem to hold the immediate future of Earth in our hands. Yet after us, the Earth will abide until burned into barrenness by the sun. The universe itself may abide endlessly, gently cooling into subatomic night.

Unanswered Questions:

1. Are current world policies leading to a sustainable future or to some kind of collapse?

2. Can new technologies alter the long-range tendency of world systems to grow and collapse?

3. Can the market system allocate resources in a sustainable way? It seems to allocate wealth to the wealthy and increase the poverty of the poor. What can change this structural element of the present system, without which it seems impossible to stabilize population growth?

4. Can industrialized people learn to live in harmony with nature? Can they share with less industrialized people?

Notes

1. Expanding into Universe

1. For this chapter I relied most heavily on Terence Dickinson, 1992, *The Universe and Beyond*, revised and expanded, Buffalo, NY: Camden; Timothy Ferris, 1997, *The Whole Shebang: A State-of-the-Universe Report*, New York: Simon and Schuster; Brian Greene, 1999, *The Elegant Universe*, New York: Vintage; and Robert T. Kirshner, 2003, *The Extravagant Universe: Exploding Stars, Dark Energy and the Accelerating Cosmos*, Princeton, NJ: Princeton University Press.
2. Bill Bryson, 2003, *A Short History of Nearly Everything*, New York: Broadway Books, 37.
3. Ibid., 12.
4. Ibid., 24.
5. Lee Smolin, 1998, *The Life of the Cosmos*, London: Phoenix.

2. Living Earth

1. See two books by James Lovelock: 1979, *Gaia: A New Look at Life on Earth*, Oxford: Oxford University Press, and 1988, *The Ages of Gaia: A Biography of our Living Earth*, New York: Bantam Books.
2. Lynn Margulis and Dorion Sagan, 1986, *Microcosmos: Four Billion Years of Evolution from Our Microbial Ancestors*, Berkeley, CA: University of California Press, new ed. 1997, 183–184.
3. For this section and the next I relied on Margulis and Sagan, 1986; Bill Bryson, 2003, *A Short History of Nearly Everything*, New York: Broadway Books; Richard Dawkins, 2004, *The Ancestor's Tale: A Pilgrimage to the Dawn of Evolution*, Boston: Houghton Mifflin.
4. Margulis and Sagan, 1986, 121–123.
5. In addition to the books named in previous notes, I used Stephen J. Gould, ed.,

1993, *The Book of Life: An Illustrated History of the Evolution of Life on Earth*, New York: W. W. Norton.

6. Bryson, 2003, ch. 14.

7. J. R. McNeill, 2000, *Something New Under the Sun: An Environmental History of the Twentieth Century World*, New York: W.W. Norton, 49.

8. Margulis and Sagan, 1986, 167.

9. Bryson, 2003, 342.

10. Douglas H. Erwin, 2006, *Extinction: How Life on Earth Nearly Ended 250 Million Years Ago*, Princeton: Princeton University Press, 10, 87.

11. Ibid., 41, 187.

12. For the fascinating story of finding the crater, see Walter Alvarez, 1997, *T. Rex and the Crater of Doom*, Princeton, NJ: Princeton University Press. Also see Bryson, 2003, ch. 13.

13. My favorite books about apes are Frans De Waal, 2005, *Our Inner Ape: A Leading Primatologist Explains Why We Are Who We Are*, New York: Riverhead Books; and Jared Diamond, 1991, *The Rise and Fall of the Third Chimpanzee*, London: Radius. Other fine books are: Roger Fouts with S.T. Mills, 1997, *Next of Kin: My Conversations with Chimpanzees*, New York: Avon; Jonathan Marks, 2002, *What It Means to Be 98% Chimpanzee: Apes, People, and Their Genes*, Berkeley, CA: University of California Press; Robert Sapolsky, 2001, *A Primate's Memoir: A Neuroscientist's Unconventional Life Among the Baboons*, New York: Simon and Schuster; and Craig Stanford, 2001, *Significant Others: The Ape-Human Continuum and the Quest for Human Nature*, New York: Basic Books.

14. Chimps are more closely related to us than they are to gorillas. If classification were based on genetic distance, than humans would belong to the same genus (*Homo*) as the other two species of chimpanzee, i.e., we would be the third species of chimpanzee. See Diamond, 1991, 20–21.

15. De Waal, 2005, 7–19. De Waal believes that the sisterhood among bonobos would not be possible without predictable, abundant food supplies (228).

16. For a clear description of the process, see Dawkins, 2004, 517–523.

17. Bryson, 2003, 308–310.

18. Stephen Pinker, 1997, *How the Mind Works*, New York: W. W. Norton, 386–389.

3. Human Emergence: One Species

1. David Christian, 2004, *Maps of Time: An Introduction to Big History*, Berkeley, CA: University of California Press, 502–503.

2. Bill Bryson, 2003, *A Short History of Nearly Everything*, New York: Broadway Books, 336.

3. My main sources for this chapter are: Richard Dawkins, 2004, *The Ancestor's Tale: A Pilgrimage to the Dawn of Evolution*, Boston: Houghton Mifflin; Brian Fagan, 1990, *The Journey from Eden: The Peopling of Our World*, London: Thames and Hudson; Stephen Jay Gould, ed., 1993, *The Book of Life: An Illustrated History of the Evolution of Life on Earth*, New York: W. W. Norton; Roger Lewin, 1988, *In the Age of Mankind*, Washington, DC: Smithsonian Books; Colin Tudge, 1996, *The Time Before History: Five Million Years of Human Impact*, New York: Scribner; and Jared Diamond, 1991, *The Rise and Fall of the Third Chimpanzee*, London: Radius.

4. See Richard W. Wrangham, 2001, "Out of the *Pan*, into the Fire: How Our Ancestors' Evolution Depended on What They Ate," in Frans B. M. De Waal, *Tree of Origin: What Primate Behavior Can Tell Us About Human Social Evolution*, Cambridge, MA: Harvard University Press, 121–143.

5. Jonathan Marks, 2002, *What It Means to Be 98% Chimpanzee: Apes, People, and Their Genes*, Berkeley, CA: University of California Press, 225.

6. Eugenia Shanklin, 1994, *Anthropology and Race*, Belmont, CA: Wadsworth, 10–12.

7. Derek Bickerton, 1995, *Language and Human Behavior*, Seattle, WA: University of Washington Press, 70.

8. Roger Fouts, 1997, *Next of Kin: My Conversations with Chimpanzees*, New York: Avon, ch. 8; and Tudge, 1996, 246–250.

9. Fagan, 1990, 19–22.

10. George Gallup Jr., and D. Michael Lindsay, 1999, *Surveying the Religious Landscape: Trends in U.S. Beliefs*, Harrisburg, PA: Morehouse Publishing, 36–37.

4. Advanced Hunting and Gathering

1. For this chapter, the following books were indispensable: Roger Lewin, 1999, *Human Evolution: An Illustrated Introduction*, 4th ed., Malden, MA: Blackwell Science; Clive Ponting, 1991, *A Green History of the World: The Environment and the Collapse of Great Civilizations*, New York: Penguin; Brian Fagan, 1992, *People of the Earth: An Introduction to World Prehistory*, 7th ed., HarperCollins; and L.S. Stavrianos, 1989, *Lifelines from Our Past: A New World History*, New York: Pantheon Books. For summarizing this material I am guided here, and in subsequent chapters, by two basic books: Richard W. Bulliet, et al., 2003, *The Earth and Its People: A Global History*, brief 2nd ed., Boston: Houghton Mifflin, and J.R. McNeill and William H. McNeill, 2003, *The Human Web: A Bird's-Eye View of World History*, New York: W. W. Norton.

2. Geoffrey Blainey, 2002, *A Short History of the World*, Chicago: Ivan R. Dee, 12.

3. Margaret Ehrenberg, 1989, *Women in Prehistory*, Norman: University of Oklahoma Press, 15–18; Fagan, 1992, 66–69.

4. Ehrenberg, 1989, ch. 2.

5. William H. McNeill, "Secrets of the Cave Paintings," *New York Review of Books*, October 19, 2006, 20–23.

6. Randall White, 1986, *Dark Caves, Bright Visions*, New York: American Museum of Natural History, 113.

7. See Riane Eisler, 1987, *The Chalice and the Blade: Our History, Our Future*, New York: HarperCollins.

8. Ehrenberg, 1989, 66–76.

9. See Derek Bickerton, 1995, *Language and Human Behavior*, Seattle, WA: University of Washington Press; Jared Diamond, 1991, *The Rise and Fall of the Third Chimpanzee*, London: Radius; and Stephen Pinker, 1994, *The Language Instinct: How the Mind Creates Language*, New York: William Morrow, 2000, HarperCollins Perennial.

10. Diamond, 1991, ch. 2, calls this the "Great Leap Forward." See also Stephen Mithen, 1996, *The Prehistory of the Mind: The Cognitive Origins of Art, Religion, and Science*, London: Thames and Hudson.

11. David Christian, 2004, *Maps of Time: An Introduction to Big History*, Berkeley, CA: University of California Press, 178–180; also Bickerton, 1995; Pinker, 1994.

12. Nicolas Wade, "In Click Languages, An Echo of the Tongues of the Ancients," *New York Times*, March 18, 2003, science section; this refers to the work of Alec Knight and Joanna Mountain.

13. From the work of Russian prehistorian Boris Frolov, reported in Ian Wilson, 2001, *Past Lives: Unlocking the Secrets of our Ancestors*, London: Cassell, 28.

14. Luigi Luca Cavalli-Sforza, 2000, *Genes, Peoples, and Languages*, New York: North Point/Farrar, Straus and Giroux, 140; Merritt Ruhlen, 1994, *The Origin of Language: Tracing the Evolution of the Mother Tongue*, New York: John Wiley, 119.

15. For discussions of human universals see Donald E. Brown, 1991, *Human Universals*, Philadelphia, PA: Temple University Press; and Mark Ridley, 1996, *The Origin of Virtue: Human Instincts and the Evolution of Cooperation*, New York: Viking Penguin.

16. For a clear account, see Loyal Rue, 2000, *Everybody's Story: Wising Up to the Epic of Evolution*, Albany, NY: State University of New York Press, 81–96.

17. See William Ryan and Walter Pitman, 1999, *Noah's Flood: The New Scientific Discoveries About the Event That Changed History*, New York: Simon and Schuster.

18. Colin Tudge, 1999, *Neanderthals, Bandits and Farming: How Agriculture Really Began*, New Haven: Yale University Press.

19. For this discussion, see Ivan Hannaford, 1996, *Race: The History of an Idea in the West*, Baltimore, MD: Johns Hopkins University Press.

20. Spencie Love, 1996, *One Blood: The Death and Resurrection of Charles R. Drew*, Chapel Hill, NC: University of North Carolina Press; John H. Relethford, 1994, *The Human Species: An Introduction to Biological Anthropology*, 2nd ed., Mountain View, CA: Mayfield Publishing, 164–173.

21. Richard Dawkins, 2004, *The Ancestor's Tale: A Pilgrimage to the Dawn of Evolution*, Boston: Houghton Mifflin, 405.

22. Relethford, 1994, 173–178.

23. Diamond, 1991, ch. 6.

24. Ibid., chs. 5 and 6.

25. Marshall Sahlins, 1972, *Stone Age Economics*, Chicago and New York: Aldine-Atherton.

26. Ridley, 1996, 6.

27. Dawkins, 2004, 32–33; Relethford, 1994, 160–161.

28. Bryan Sykes, 2001, *The Seven Daughters of Eve: The Science That Reveals Our Genetic Ancestry*, New York: W. W. Norton.

5. Early Agriculture

1. This chapter relies heavily on J.R. McNeill and William H. McNeill, 2003, *The Human Web: A Bird's-Eye View of World History*, New York: W. W. Norton; Richard W. Bulliet, et al., 2003, *The Earth and Its People: A Global History*, brief 2nd ed., Boston: Houghton Mifflin; and John A. Mears, 2001, "Agricultural Origins in Global Perspective," in Michael Adas, ed., *Agricultural and Pastoral Societies in Ancient and Classical History*, Philadelphia: Temple University Press, 36–70.

2. See Mark Cohen, 1977, *The Food Crisis in Prehistory: Overpopulation and the Origins of Agriculture*, New Haven: Yale University Press; and Stephen Jay Gould,

"Down on the Farm, A review of Donald O. Henry, *From Foraging to Agriculture: The Levant at the End of the Ice Age,*" *New York Review of Books,* January 18, 1990, 26–27.

3. Charles B. Heiser, 1981, *Seed to Civilization: The Story of Food,* 2nd ed., San Francisco: W. H. Freeman, 16; Stephen Budiansky, 1992, *The Covenant of the Wild: Why Animals Chose Domestication,* New York: William Morrow, 82.

4. Budiansky, 1992, 22.

5. Margaret Ehrenberg, 1989, *Women in Prehistory,* Norman: University of Oklahoma Press, 77–78; Riane Eisler, 1987, *The Chalice and the Blade: Our History, Our Future,* New York: HarperCollins, 68–69.

6. David Christian, 2004, *Maps of Time: An Introduction to Big History,* Berkeley: University of California Press, 208; Vaclav Smil, 1994, *Energy in World History,* Boulder, CO: Westview Press, 236.

7. Jared Diamond, 1999, *Guns, Germs and Steel: The Fates of Human Societies,* New York: Norton.

8. L.S. Stavrianos, 1989, *Lifelines from Our Past: A New World History,* New York: Pantheon Books, 48.

9. James Mellaart, 1967, *Çatal Hüyük: A Neolithic Town in Anatolia,* New York: McGraw-Hill; Ian A. Todd, 1976, *Çatal Hüyük in Perspective,* Menlo Park, CA: Cummings Publishing; Eisler, 1987.

10. Malcolm Gladwell, 2000, *The Tipping Point: How Little Things Can Make a Big Difference,* Boston: Bay Books, Little, Brown, 178–180; Fred Spier, 1996, *The Structure of Big History: From the Big Bang Until Today,* Amsterdam: Amsterdam University Press, 62–66.

11. John Noble Wilford, 1993, "9000-Year-old Cloth Found," *San Francisco Chronicle,* August 13, 1993.

12. Mark Kurlansky, 2002, *Salt: A World History,* New York: Walker, 6–12.

13. Heiser, 1981, 20–22; Catherine Johns, 1982, *Sex or Symbol? Erotic Images of Greece and Rome,* London: British Museum, 39–40.

14. Sarah Shaver Hughes and Brady Hughes, 2001, "Women in Ancient Civilizations," in Michael Adas, ed., *Agricultural and Pastoral Societies of Ancient and Classical History,* Philadelphia, PA: Temple University Press, 116–150; Donald E. Brown, 1991, *Human Universals,* Philadelphia: Temple University Press, 52.

15. Clive Ponting, 1991, *A Green History of the World: The Environment and the Collapse of Great Civilizations,* New York, Penguin; Roger Sands, 2005, *Forestry in a Global Context,* Cambridge, MA: CABI Publishing.

16. Quoted in Daniel J. Hillel, 1991, *Out of the Earth: Civilization and the Life of the Soil,* New York: Free Press, Macmillan, 16.

17. J. R. McNeill, 2005, *Something New Under the Sun: An Environmental History of the Twentieth Century World,* New York: W. W. Norton, 45.

18. Stephen Mitchell, 2004, *Gilgamesh: A New English Version,* New York: Free Press.

19. Hillel, 1991, 63, agrees with this interpretation, as does Colin Tudge, 1996, *The Time Before History: Five Million Years of Human Impact,* New York: Scribner, 267.

20. Michael Pollan, 2001, *The Botany of Desire: A Plant's Eye View of the World,* New York: Random House, 11.

21. For further reading see William Ryan and Walter Pitman, 1999, *Noah's Flood: The New Scientific Discoveries About the Event that Changed History,* New York: Simon and Schuster.

6. Early Cities

1. For one discussion see Peter N. Stearns, 1987, *World History: Patterns of Change and Continuity*, New York: Harper and Row, 13–16.
2. David Christian, 2004, *Maps of Time: An Introduction to Big History*, Berkeley: University of California Press, 248.
3. For this chapter my basic sources are J.R. McNeill and William H. McNeill, 2003, *The Human Web: A Bird's-Eye View of World History*, New York: W. W. Norton; Richard W. Bulliet, et al., 2003, *The Earth and Its People: A Global History*, brief 2nd ed., Boston: Houghton Mifflin; and Arthur Cotterell, ed., 1980, *The Penguin Encyclopedia of Ancient Civilizations*, London and New York: Penguin.
4. About the Sumerians, see Harriet Crawford, 1991, *Sumer and the Sumerians*, Cambridge: Cambridge University Press.
5. Christian, 2004, 261.
6. Rosalind Miles, 1990, *The Women's History of the World*, New York: Harper and Row, 43.
7. Crawford, 1991, 151–153; Georges Jean, 1992, *Writing: The Story of Alphabets and Scripts*, New York: Harry N. Abrams, 12–21.
8. Dale Keiger, "Clay, Paper, Code," *Johns Hopkins Magazine*, September 2003, 34–41; Timothy Potts, "Buried Between the Rivers," *New York Review of Books*, September 25, 2003, 18–23; Diane Wolkstein and Samuel Noah Kramer, 1983, *Inanna: Queen of Heaven and Earth: Her Stories and Hymns from Sumer*, New York: Harper and Row, 127–135.
9. Brian M. Fagan, 2005, *The Long Summer: How Climate Changed Civilization*, New York: Basic Books, 6–7, 141–145.
10. Jared Diamond, 1999, *Guns, Germs and Steel: The Fates of Human Societies*, New York: Norton, 418.
11. Wolkstein and Kramer, 1983, 101.
12. Charles Officer and Jake Page, 1993, *Tales of the Earth: Paroxysms and Perturbations of the Blue Planet*, New York: Oxford University Press, 62–63.
13. Mark Kurlansky, 2002, *Salt: A World History*, New York: Walker, 38–44.
14. Daniel J. Hillel, 1991, *Out of Earth: Civilization and the Life of the Soil*, New York: Free Press, Macmillan, 5.
15. R. F. Willetts, 1980, "The Minoans," in Arthur Cotterell, ed., *The Penguin Encyclopedia of Ancient Civilizations*, London and New York: Penguin, 204–210; William J. Broad, "It Swallowed a Civilization," *New York Times*, October 21, 2003, E1–2.
16. Miles, 1990, 64.
17. Colin A. Ronan, 1978, *The Shorter Science and Civilization in China: An Abridgement of Joseph Needham's Original Text*, vol. I, Cambridge, England: Cambridge University Press, 26–30.
18. Kurlansky, 2002, 44–46; Jerry H. Bentley, 1993, *Old World Encounters: Cross Cultural Contacts and Exchanges in Pre-Modern Times*, New York: Oxford University Press.
19. Hillel, 1991, 16–17.
20. Jean, 1987, 60–62.
21. Christian, 2004, 257, 263.
22. Joseph Campbell with Bill Moyers, 1988, *The Power of Myth*, New York: Doubleday, 169–171.
23. Miles, 1990, 47.

24. Catherine Johns, 1982, *Sex or Symbol? Erotic Images of Greece and Rome,* London: British Museum Press, 42–61.
25. Christian, 2004, 143, 309.
26. Ibid., 258; Geoffrey Blainey, 2002, *A Short History of the World,* Chicago: Ivan R. Dee, 72–73.
27. Cotterell, ed., 1980, 16–17; Bentley, 1993, 21.
28. See a review of the video *Black Athena* by Franklin W. Knight, 1993, "*Black Athena,*" *Journal of World History* 4, no. 2, Fall 1993, 325–327.

7. The Afro-Eurasian Network

1. Mark Kurlansky, 2002, *Salt: A World History,* New York: Walker, 54–55.
2. Peter Berresford Ellis, 1990, *The Celtic Empire: The First Millennium of Celtic History, c. 1000 B.C.–51 A.D.,* London: Constable.
3. Richard W. Bulliet, et al., 2003, *The Earth and Its People: A Global History,* brief 2nd ed., Boston: Houghton Mifflin, 108–115; J.R. McNeill and William H. McNeill, 2003, *The Human Web: A Bird's-Eye View of World History,* New York: W. W. Norton, 67–68; 86.
4. See Huston Smith and Phil Novak, 2003, *Buddhism: A Concise Introduction,* New York: HarperCollins.
5. Robert Temple, 1986, *The Genius of China: Three Thousand Years of Science, Discovery and Invention,* New York: Simon and Schuster, 219–224.
6. McNeill and McNeill, 2003, 67.
7. Xinru Liu, 2001, "The Silk Road: Overland Trade and Cultural Interactions in Eurasia," in Michael Adas, ed., *Agricultural and Pastoral Societies in Ancient and Classical History,* Philadelphia: Temple University Press, 151–179.
8. Vaclav Smil, 1994, *Energy in World History,* Boulder, CO: Westview Press, 232.
9. McNeill and McNeill, 2003, 80.
10. Crane Brinton, John B. Christopher, and Robert Lee Wolff, 1960, *A History of Civilization,* 2nd ed., Englewood Cliffs, NJ: Prentice-Hall, I, 65.
11. Sarah Shaver Hughes and Brady Hughes, 2001, "Women in Ancient Civilizations," in Michael Adas, ed., *Agricultural and Pastoral Societies in Ancient and Classical History,* Philadelphia: Temple University Press, 140.
12. Michael Cook, 2003, *A Brief History of the Human Race,* New York: W. W. Norton, 226.
13. Donald J. Hughes, 1975, *Ecology in Ancient Civilizations,* Albuquerque: University of New Mexico, 68–75.
14. Kurlansky, 2002, 63–68.
15. Shaye J. D. Cohen, 1988, "Roman Domination: The Jewish Revolt and the Destruction of the Second Temple," in Hershel Shanks, ed., *Ancient Israel: A Short History from Abraham to the Roman Destruction of the Temple,* Englewood Cliffs, NJ: Prentice-Hall, 205–235.
16. William H. McNeill, 1976, *Plagues and People,* Garden City, NJ: Anchor Press/Doubleday, 121–122; Robert Austin Markus, 1974, *Christianity in the Roman World,* London: Thames and Hudson, 25.
17. William H. McNeill, 1992, *The Global Condition: Conquerors, Catastrophes and Community,* Princeton: Princeton University Press, 103.

18. David Christian, 2004, *Maps of Time: An Introduction to Big History*, Berkeley: University of California Press, 143, 325–326.

19. Clive Ponting, 1991, *A Green History of the World: The Environment and the Collapse of Great Civilizations*, New York: Penguin, 54–83; Hughes, 1975, 99–124.

20. One of the earliest religions that believed in one God arose in the eastern provinces of the Persian empire by the beginning of the fifth century BCE. Called Zoroastrianism, its prophet was Zoroaster (Greek) or Zarathushtra (Persian), whose dates are not certain. Zoroastrians believe in a dualist universe, with forces of good and evil locked in cosmic struggle until good will prevail at the end of time. Its priests are called magi. Zoroastrianism may have influenced Judaism and Christianity. The later Islamic conquest of Iran caused the decline of Zoroastrianism; surviving communities are today called Parsees.

21. For further reading on the Axial Age, see Karen Armstrong, 2006, *The Great Transformation: The Beginning of Our Religious Traditions*, New York: Alfred A. Knopf; on why people still believe ancient religions, see Sam Harris, 2004, *The End of Faith*, New York: W. W. Norton.

22. See G. W. Bowerstock, 1988, "The Dissolution of the Roman Empire," in Norman Yoffee and George L. Cowgill, eds., *The Collapse of Ancient States and Civilizations*, Tucson, AZ: University of Arizona Press, 165–175; Allen M. Rollins, 1983, *The Fall of Rome: A Reference Guide*, Jefferson, NC: McFarland; and Bryan Ward-Perkins, 2005, *The Fall of Rome and the End of Civilization*, Oxford: Oxford University Press.

8. Expanding the Afro-Eurasian Network

1. My basic references are J.R. McNeill and William H. McNeill, 2003, *The Human Web: A Bird's-Eye View of World History*, New York: W. W. Norton, ch. IV; and Richard W. Bulliett, et al., 2003, *The Earth and Its People: A Global History*, brief 2nd ed., Boston: Houghton Mifflin, chs. 7–9.

2. Bulliett, et al., 2003, 324; see David Christian, 1998, *A History of Russia, Central Asia and Mongolia*, vol. 1, *Inner Eurasia from Prehistory to the Mongol Empire*, Oxford: Blackwell, 346–348, for a more complex discussion.

3. Crane Brinton, John B. Christopher, and Robert Lee Wolff, 1960, *A History of Civilization*, 2nd ed., Englewood Cliffs, NJ: Prentice-Hall, I, 221–222.

4. Xinru Liu, 2001, "The Silk Road: Overland Trade and Cultural Interactions in Eurasia," in Michael Adas, ed., *Agricultural and Pastoral Societies in Ancient and Classical History*, Philadelphia: Temple University Press, 154–179.

5. Richard M. Eaton, 1990, *Islamic History as Global History*, Washington, DC: American Historical Association, 10–11.

6. Frederick Kilgour, 1998, *The Evolution of the Book*, Oxford and New York: Oxford University Press, 54–62.

7. Liu, 2001, 161.

8. S.A.M. Adshead, 2000, *China in World History*, 3rd ed., New York: St. Martin's Press, 54–56.

9. Bulliet, et al., 2003, 222–223.

10. Adshead, 2000, 72–88.

11. Ibid., 70, 98.

12. Robert Temple, 1986, *The Genius of China: Three Thousand Years of Science, Discovery and Invention*, New York: Simon and Schuster, 224–228.

13. Adshead, 2000, 97.

14. Colin A. Ronan, 1978, *The Shorter Science and Civilization in China: An Abridgement of Joseph Needham's Original Text*, vol. I, Cambridge, England: Cambridge University Press, 44–48.

15. See Peter B. Golden, 2001, "Nomads and Sedentary Societies in Eurasia," in Michael Adas, ed., *Agricultural and Pastoral Societies in Ancient and Classical History*, Philadelphia: Temple University Press, 71–115.

16. Bulliet, et al., 2003, 206–207; Stewart Brand, 1999, *The Clock of the Long Now: Time and Responsibility*, New York: Basic Books, 101.

17. On the Vikings, see Gwyn Jones, 1984, *A History of the Vikings*, rev. ed., Oxford and New York: Oxford University Press; E.O.G. Turville-Petre, 1975, *Myth and Religion of the North: The Religion of Ancient Scandinavia*, Westport, CN: Greenwood Press; and David M. Wilson, 1989, *Vikings and Their Origins: Scandinavia in the First Millennium*, rev. ed., London: Thames and Hudson.

18. For a discussion of this question, see Jerry H. Bentley, 1993, *Old World Encounters: Cross Cultural Contacts and Exchanges in Pre-Modern Times*, New York: Oxford University Press, 100–101.

19. See Roger Collins, 1998, *Charlemagne*, Toronto and Buffalo: University of Toronto Press.

20. Lester Kurtz, 1995, *Gods in the Global Village: The World's Religions in Sociological Perspective*, Thousand Oaks, CA: Pine Forge Press, 271.

21. McNeill and McNeill, 2003, 98.

22. Philip D. Curtin, 1984, *Cross-Cultural Trade in World History*, Cambridge: Cambridge University Press, 15–27; John Iliffe, 1995, *Africans: The History of a Continent*, New York: Cambridge University Press, argues that African history is unique due to its obstacles—climate, geography, and diseases.

23. Curtin, 1984, 38–39; McNeill and McNeill, 2003, 96; David Christian, 2004, *Maps of Time: An Introduction to Big History*, Berkeley: University of California Press, 344.

24. David Christian, 2003, "World History in Context," *Journal of World History* 14, no. 4, 451.

25. Christian, 2004, 344.

26. Johan Goudsblom, Eric Jones, and Stephen Mennell, 1996, *The Course of Human History: Economic Growth, Social Process and Civilization*, Armonk, NY: M.E. Sharpe, 22–28.

27. Basil Davidson, 1991, *African Civilization Revisited*, Trenton, NJ: Africa World Press, 93–97.

9. Emergence of American Civilizations

1. Roger Lewin, 1988, *In the Age of Mankind*, Washington, DC: Smithsonian Books, 167–169; Jared Diamond, 1999, *Guns, Germs, and Steel: The Fates of Human Societies*, New York: Norton, 44–48.

2. Basic to this chapter is John E. Kicza, 2001, "The People and Civilizations of the Americas Before Contact," in Michael Adas, ed., *Agricultural and Pastoral Societies in Ancient and Classical History*, Philadelphia: Temple University Press, 183–223.

3. Charles C. Mann, 2005, *1491: New Revelations of the Americas Before Columbus*, New York: Alfred A. Knopf, ch. 9; William H. McNeill, "New World Symphony," *New York Review of Books*, December 1, 2005, 45.

4. J. R. McNeill and William H. McNeill, 2003, *The Human Web: A Bird's-Eye View of World History*, New York: W. W. Norton, 109; John A. Mears, 2001, "Agricultural Origins in Global Perspective," in Michael Adas, ed., *Agricultural and Pastoral Societies in Ancient and Classical History*, Philadelphia: Temple University Press, 57.

5. Kicza in Adas, ed., 2001, 185.

6. Brian M. Fagan, 2005, *The Long Summer: How Climate Changed Civilization*, New York: Basic Books, 214–230; Joseph A. Tainter, 1989, *The Collapse of Complex Societies*, Cambridge: Cambridge University Press, 170–175.

7. Brian M. Fagan, 1984, *The Aztecs*, New York: W. H. Freeman, 243–244.

8. Diamond, 1999, 292.

9. Fagan, 1984, 9–11; my account of the Aztecs is based on this book.

10. Terence N. D'Altroy, 2002, *The Incas*, Malden, MA: Blackwell, ch. 2.

11. Garcilaso de la Vega, *Royal Commentaries of the Incas*, and Felipe Guaman Pom de Ayala; see D'Altroy, 2002, 14–15. My account of the Incas is based on D'Altroy, 2002, and Craig Morris and Adriana von Hagen, 1993, *The Inka Empire and Its Andean Origins*, New York: Abbeville Press and the American Museum of Natural History.

12. Hugh Thomson, 2001, *The White Rock: An Exploration of the Inca Heartland*, Woodstock and New York: Overlook Press, 204; John Hemming, 1970, *The Conquest of the Incas*, New York: Macmillan, 498.

13. Jack Weatherford, 1991, *Native Roots: How the Indians Enriched America*, New York: Crown Publishers, 97–98.

14. Richard W. Bulliet, et al., 2003, *The Earth and Its People: A Global History*, brief 2nd ed., Boston: Houghton Mifflin, 250; McNeill and McNeill, 2003, 112.

15. Kicza, in Adas, ed., 2001, 27.

16. I. A. Ritchie Carson, 1981, *Food in Civilization: How History Has Been Affected by Human Tastes*, New York and Toronto: Beaufort Books, 106.

17. Johan Goudsblom, Eric Jones, and Stephen Mennell, 1996, *The Course of Human History: Economic Growth, Social Process and Civilization*, Armonk, NY: M.E. Sharpe.

18. McNeill and McNeill, 2003, 162.

19. Lewin, 1988, 160–164.

20. Fagan, 1984, 233–236.

21. Morris and von Hagen, 1993, 86.

10. One Afro-Eurasia

1. David Christian, 2004, *Maps of Time: An Introduction to Big History*, Berkeley: University of California Press, 305, 335.

2. Ibid., 318.

3. Jack Weatherford, 2004, *Genghis Khan and the Making of the Modern World*, New York: Crown Publishers, xxvi.

4. Ibid., introduction; David Christian, 1998, *A History of Russian, Central Asia and*

Mongolia, vol. 1, *Inner Eurasia from Prehistory to the Mongol Empire*, Oxford: Blackwell, 426.

5. David Morgan, 1986, *The Mongols*, Oxford: Basil Blackwell, 30; Weatherford, 2004, xxvii. My account of the Mongols is based on these two books, plus Peter B. Golden, 2001, "Nomads and Sedentary Societies in Eurasia," in Michael Adas, ed., *Agricultural and Pastoral Societies in Ancient and Classical Times*, Philadelphia: Temple University Press, 71–115; and Paul Ratchnevsky, 1991, *Genghis Khan: His Life and Legacy*. Oxford: Blackwell.

6. Morgan, 1986, 93; Weatherford, 2004, 113–117.

7. Quoted in Jerry H. Bentley, 1993, *Old World Encounters: Cross Cultural Contacts and Exchanges in Pre-Modern Times*, New York: Oxford University Press, 111.

8. William H. McNeill, 1976, *Plagues and People*, Garden City, NJ: Anchor Press/Doubleday, 166–186.

9. Philip D. Curtin, 1984, *Cross-Cultural Trade in World History*, Cambridge: Cambridge University Press, 125; Christian, 2004, 379; Richard W. Bulliet, et al., 2003, *The Earth and Its People: A Global History*, brief 2nd ed., Boston: Houghton Mifflin, 290–292. For a readable account, see Louise Levathes, 1994, *When China Ruled the Seas: The Treasure Fleet of the Dragon Throne, 1405–1433*, New York: Simon and Schuster.

10. Janet L. Abu-Lughod, 1989, *Before European Hegemony: The World System* A.D. *1250–1350*, New York: Oxford University Press, 344–347.

11. Curtin, 1984, 107; Bentley, 1993, 176.

12. Richard M. Eaton, 1990, *Islamic History as Global History*, Washington, DC: American Historical Association, 23.

13. J.R. McNeill and William H. McNeill, 2003, *The Human Web: A Bird's-Eye View of World History*, New York: W. W. Norton, 132.

14. Eaton, 1990, 44–45; Ross E. Dunn, 1986, *The Adventures of Ibn Battuta: A Muslim Traveler of the Fourteenth Century*, Berkeley, University of California Press.

15. Bulliet, et al., 2003, 319–320; Paul E. Lovejoy, 2000, *Transformations in Slavery: A History of Slavery in Africa*, 2nd ed., Cambridge: Cambridge University Press, 24–25.

16. L.S. Stavrianos, 1989, *Lifelines from Our Past: A New World History*, New York: Pantheon, 54.

17. See Christopher Tyerman, 2006, *God's War: A New History of the Crusades*, Cambridge, MA: Harvard University Press.

18. Morgan, 1986, 179; Weatherford, 2004, 162.

19. Roger Sands, 2005, *Forestry in a Global Context*, Cambridge, MA: CABI Publishing, 31–33.

20. Frederick Kilgour, 1998, *The Evolution of the Book*, New York: Oxford University Press, 8, 82.

21. Christian, 2004, 344–345.

22. See the debate in the *Journal of World History* 4, no. 4, December 2003, 503–550.

23. See Nicolas Wade, "A Prolific Genghis Khan, It Seems, Helped People the World," *New York Times*, February 11, 2003, D3.

24. See Robert Finlay, "How Not to (Re)Write World History: Gavin Menzies and the Chinese Discovery of America," *Journal of World History* 15, 2, June 2004, 229–242. For a sound account, see Levathes, 1994.

11. Connecting the Globe

1. Richard W. Bulliet, et al., 2003, *The Earth and Its People: A Global History,* brief 2nd ed., Boston: Houghton Mifflin, 344; J.R. McNeill and William H. McNeill, 2003, *The Human Web: A Bird's-Eye View of World History,* New York: W. W. Norton, 163, 176.

2. David Christian, 2004, *Maps of Time: An Introduction to Big History*, Berkeley: University of California Press, 381.

3. John A. Mears, 2001, "Agricultural Origins in Global Perspective," in Michael Adas, ed., *Agricultural and Pastoral Societies in Ancient and Classical History*, Philadelphia: Temple University Press, 38; Jared Diamond, 1999, *Guns, Germs and Steel: The Fates of Human Societies*, New York: Norton, 266, 283; Christian, 2004, 344–345.

4. McNeill and McNeill, 2003, 156–161.

5. Jerry H. Bentley, 1993, *Old World Encounters: Cross Cultural Contacts and Exchanges in Pre-Modern Times*, New York: Oxford University Press, 177.

6. James Reston, 2005, *Dogs of God: Columbus, the Inquisition, and the Defeat of the Moors*, New York: Doubleday, 205. England had expelled its Jews in 1290, Ibid.

7. Ivan Hannaford, 1996, *Race: The History of an Idea in the West.* Washington, DC: Woodrow Wilson Center Press, ch. 4; George M. Frederickson, 2002, *Racism: A Short History*, Princeton: Princeton University Press, 17–34.

8. Milton Meltzer, 1990, *Columbus and the World Around Him*, New York: Franklin Watts.

9. On Cortez and the Aztecs, see Brian M. Fagan, 1984, *The Aztecs*, New York: W. H. Freeman, ch. 11.

10. Fagan, 1984, ch. 12.

11. On the conquest of Peru, see Diamond, 1999, 68–79; Terence N. D' Altroy, 2002, *The Incas*, Malden, MA: Blackwell; Nigel Davies, 1995, *The Incas*, Niwot, CO: University Press of Colorado; John Hemming, 1970, *The Conquest of the Incas*, New York: Macmillan.

12. Diamond, 1999.

13. William H. McNeill, 1976, *Plagues and People*, Garden City, NJ: Anchor Press/Doubleday; Robert S. Desowitz, 1997, *Who Gave Pinta to the Santa Maria? Tracking the Devastating Spread of Lethal Tropical Disease into America*, New York: Harcourt Brace.

14. Hemming, 1970, 267–288; Hugh Thomas, 2004, *Rivers of Gold: The Rise of the Spanish Empire, from Columbus to Magellan*, New York: Random House, 304–456.

15. On gold and silver, see Jack Weatherford, 1988, *Indian Givers: How the Indians of the Americas Transformed the World*, New York: Fawcett Columbine, 6–17.

16. I.A. Ritchie Carson, 1981, *Food in Civilization: How History Has Been Affected by Human Tastes*, New York and Toronto: Beaufort Books, 111–128; J.M. Blaut, 1993, *The Colonizer's Model of the World: Geographical Diffusionism and Eurocentric History*, New York and London: Guilford Press, 191–192; Bulliet, et al., 2003, 398.

17. Frederickson, 2002, 30; McNeill and McNeill, 2003, 168–169; William H. McNeill, 1992, *The Global Condition: Conquerors, Catastrophes and Community*, Princeton: Princeton University Press, 21.

18. Paul Bairoch, 1993, *Economics and World History: Myths and Paradoxes*, Chicago: University of Chicago Press, 146–147; Patrick Manning, 1990, *Slavery and African Life: Occidental, Oriental and African Trade*, Cambridge and New York: Cambridge University Press, 171; Bulliet, et al., 2003, 421.

19. Fernand Braudel, 1985, *Civilizations and Capitalism, Fifteenth–Eighteenth Century*, vol. II, London: Fontana Press, 101–102.
20. Ibid., 559–568.
21. This section relies heavily on McNeill and McNeill, 2003, 186–212.
22. This phrase comes from Arnold Pacey, 1990, *Technology in Civilization: A Thousand-Year History*, Cambridge, MA: MIT Press, 62.
23. McNeill and McNeill, 2003, 170–171.
24. Manning, 1990, 84.
25. Bulliet, et al., 2003, 384; Thomas, 2004, 304–456; Frederickson, 2002, 36.

12. Industrialization

1. Peter N. Stearns, 1993, *The Industrial Revolution in World History*, Boulder, CO: Westview Press. For example, Fred Spier, 1996, *The Structure of Big History: From the Big Bang Until Today*, Amsterdam: Amsterdam University Press, 38.
2. J.R. McNeill and William H. McNeill, 2003, *The Human Web: A Bird's-Eye View of World History*, New York: W. W. Norton, 222; Donella Meadows, et al., 2004, *Limits to Growth: The Thirty-Year Update*, White River Junction, VT: Chelsea Green Publishing, 28.
3. Robert Skidelsky, "The Mystery of Growth," *New York Review of Books*, March 13, 2003, 28–31.
4. Fernand Braudel, 1985, *Civilizations and Capitalism, Fifteenth–Eighteenth Century*, vol. II, London: Fontana Press, 245, 525–574.
5. Charles Van Doren, 1991, *A History of Knowledge: Past, Present and Future*, New York: Ballantine, 227.
6. Howard Zinn, 1980, *A People's History of the United States*, New York: Harper and Row, 82–95.
7. George M. Frederickson, 2002, *Racism: A Short History*, Princeton: Princeton University Press, 56–57; Ivan Hannaford, 1996, *Race: The History of an Idea in the West*, Washington, DC: Woodrow Wilson Center Press, 206–208.
8. Kenneth Pomeranz, 2001, *The Great Divergence: China, Europe, and the Making of the Modern World Economy*, Princeton: Princeton University Press, 61; J.R. McNeill, 2000, *Something New Under the Sun: An Environmental History of the Twentieth-Century World*, New York: W. W. Norton, 13; McNeill and McNeill, 2003, 230–232.
9. Pomeranz, 2001, 61–66.
10. Crane Brinton, John B. Christopher, and Robert Lee Wolff, 1960, *A History of Civilization*, vol. II, Englewood Cliffs, NJ: Prentice-Hall, 11–12; I.A. Ritchie Carson, 1981, *Food in Civilization: How History Has Been Affected by Human Tastes*, New York and Toronto: Beaufort Books, 135–136.
11. Braudel, 1985, vol. III, 595–615; Paul Kennedy, 1991, *Preparing for the Twenty-First Century*, New York: Ballantine, 6–7.
12. Braudel, 1985, vol. I, 249–261; Pomeranz, 2001, 275–281.
13. Louise A. Tilly, 1993, *Industrialization and Gender Inequality*, Washington, DC: American Historical Association, 14, 48; Stearns, 1991, 14–15.
14. McNeill and McNeill, 2003, 241–245; Stearns, 1991, 35–40.
15. David Christian, 1997, *Imperial and Soviet Russia: Power, Privilege and the Chal-

lenge of Modernity, New York: St. Martin's Press, ch. 3; McNeill and McNeill, 2003, 252–258.

16. This paragraph and the previous one are based on McNeill and McNeill, 2003, 217–221.

17. J.R. McNeill, 2000, 14; Kennedy, 1991, 32.

18. Philip Curtin, 1984, *Cross-Cultural Trade in World History*, Cambridge: Cambridge University Press, 251; Peter Jay, 2000, *The Wealth of Man*, New York: Public Affairs, 186–208.

19. Frederickson, 2000, 170.

20. Paul E. Lovejoy, 2000, *Transformations in Slavery: A History of Slavery in Africa*, 2nd ed., Cambridge: Cambridge University Press, 252, 288; McNeill and McNeill, 2003, 216; John Iliffe, 1995, *Africans: The History of a Continent*, New York: Cambridge University Press, 3.

21. See Mike Davis, 2001, *Late Victorian Holocausts: El Niño Famines and the Making of the Third World*, London and New York: Verso. Also John Richards, 2003, *The Unending Frontier: An Environmental History of the Early Modern World*, Berkeley: University of California Press, 82.

22. McNeill and McNeill, 2003, 291–292; Paul Bairoch, 1993, *Economics and World History: Myths and Paradoxes*, Chicago: University of Chicago Press, 9.

23. McNeill and McNeill, 2003, 280.

24. Lester Kurtz, 1995, *Gods in the Global Village: The World's Religions in Sociological Perspective*, Thousand Oaks, CA: Pine Forge Press, 21.

25. J.R. McNeill, 2000, 15–16; McNeill and McNeill, 2003, 290–316.

26. Kennedy, 1991, 45–46; Donella Meadows, Jorgen Randers, and Dennis Meadows, 2004, *Limits to Growth: The 30-Year Update*, White River Junction, VT: Chelsea Green, 42–43.

27. David Christian, 2004, *Maps of Time: An Introduction to Big History*, Berkeley: University of California Press, 444–445; Paul Kivel, 2004, *You Call This a Democracy? Who Benefits, Who Pays and Who Really Decides*, New York: Apex Press, 25.

28. J.R. McNeill, 2000, 16.

29. Braudel, 1985, vol. III, 620–623.

30. Allen K. Smith, 1991, *Creating a World Economy: Merchant Capital, Colonialism and World Trade, 1400–1825*, Boulder, CO: Westview Press, 116.

31. Elizabeth Kolbert, "Why Work?" *New Yorker*, November 29, 2004, 154.

13. What Now? What Next?

1. J.R. McNeill, 2000, *Something New Under the Sun: An Environmental History of the Twentieth Century World*, New York: W. W. Norton, 9. The rest of this section is based on Bjorn Lomborg, 2001, *The Skeptical Environmentalist: Measuring the Real State of the World*, Cambridge: Cambridge University Press.

2. Donella Meadows, et al., 1972, *The Limits to Growth: A Report for the Club of Rome's Project on the Predicament of Mankind*, New York: Universe Books.

3. This section is based on J.R. McNeill, 2000, unless otherwise noted.

4. J.R. McNeill, 2000, 111–115; David Christian, 2004, *Maps of Time: An Introduction to Big History*, Berkeley: University of California Press, 478–479.

5. J.R. McNeill, 2000, 108–111; Jim Hansen, "The Threat to the Planet," *New York Review of Books*, July 13, 2006, 12–14, 16.

6. Kenneth Pomeranz, 2001, *The Great Divergence: China, Europe, and the Making of the Modern World Economy*, Princeton: Princeton University Press, 56–60; Donella Meadows, Jorgen Randers, and Dennis Meadows, 2004, *Limits to Growth: The 30-Year Update*, White River Junction, VT: Chelsea Green, 75.

7. J.R. McNeill, 2000, 229; Jared Diamond, 2005, *Collapse: How Societies Choose to Fail or Succeed*, New York: Viking Penguin, 473, 487.

8. J.R. McNeill, 2000, 146; Meadows, Randers, and Meadows, 2004, 229–231.

9. Meadows, Randers, and Meadows, 2004, 66–74.

10. J.R. McNeill, 2000, 342; Martin Rees, 2003, *Our Final Hour: A Scientist's Warning: How Terror, Error, and Environmental Disaster Threatens Humankind's Future in this Century—On Earth and Beyond*, New York: Basic Books, 34–35; James Sterngold, "Experts Fear Nuke Genie's Out of Bottle," *San Francisco Chronicle*, November 22, 2004, A8.

11. J.R. McNeill, 2000, 312; Lomborg, 2001, 129.

12. Lomborg, 2001, 129.

13. See Richard Leakey and Roger Lewin, 1995, *The Sixth Extinction: Patterns of Life and the Future of Mankind*, New York: Doubleday.

14. Meadows, Randers, and Meadows, 2004, xiv–xv.

15. Stephen Moore and Julian L. Simon, 2000, *It's Getting Better All the Time: 100 Greatest Trends of the Last 100 Years*, Washington, DC: Cato Institute, 23.

16. Meadows, Randers, and Meadows, 2004, ch. 5; Christian, 2004, 478–479.

17. Quoted by Meadows, Randers, and Meadows, 2004, 13.

18. Ibid., 15.

19. Nikos Prantzos, 2000, *Our Cosmic Future: Humanity's Fate in the Universe*, Cambridge: Cambridge University Press, 10, 18, 56–85.

20. For the long-range future, see Prantzos, 2000, and Christian, 2004, 486–490.

BIBLIOGRAPHY

This bibliography provides complete details of all the books and articles referred to in the text or captions, plus a few additional books too important or wonderful to omit.

Abu-Lughod, Janet L. 1989. *Before European Hegemony: The World System A.D. 1250–1350.* New York and Oxford: Oxford University Press.

Adas, Michael, ed. 2001. *Agricultural and Pastoral Societies in Ancient and Classical History.* Philadelphia: Temple University Press.

Adshead, S.A.M. 2000. *China in World History.* 3rd ed. New York: St. Martin's Press.

Alvarez, Walter. 1997. *T. Rex and the Crater of Doom.* Princeton, NJ: Princeton University Press.

Anderson, Walter Truett. 2001. *All Connected Now: Life in the First Global Civilization.* Boulder, CO: Westview Press.

Armstrong, Karen. 2006. *The Great Transformation: The Beginning of Our Religious Traditions.* New York: Alfred A. Knopf.

Bairoch, Paul. 1993. *Economics and World History: Myths and Paradoxes.* Chicago: University of Chicago Press.

Bakker, Robert. 1986. *The Dinosaur Heresies: New Theories Unlocking the Mystery of the Dinosaurs and Their Extinctions.* New York: William Morrow.

Bentley, Jerry H. 1993. *Old World Encounters: Cross Cultural Contacts and Exchanges in Pre-Modern Times.* New York and Oxford: Oxford University Press.

Bernal, Martin. 1987. *Black Athena: The Afroasiatic Roots of Classical Civilization. Vol. 1: The Fabrication of Ancient Greece, 1785–1985.* New Brunswick, NJ: Rutgers University Press.

Bickerton, Derek. 1995. *Language and Human Behavior.* Seattle: University of Washington Press.

Blainey, Geoffrey. 2002. *A Short History of the World.* Chicago: Ivan R. Dee.

Blaut, J.M. 1993. *The Colonizer's Model of the World: Geographical Diffusionism and Eurocentric History.* New York and London: Guilford Press.

Bowerstock, G.W. 1988. "The Dissolution of the Roman Empire." In Norman Yoffee and

George L. Cowgill, eds. *The Collapse of Ancient States and Civilizations*. Tucson: University of Arizona Press.

Brand, Stewart. 1999. *The Clock of the Long Now: Time and Responsibility*. New York: Basic Books.

Braudel, Fernand. 1985. *Civilizations and Capitalism, Fifteenth–Eighteenth Century*. 3 vols. London: Fontana Press.

Brinton, Crane, John B. Christopher, and Robert Lee Wolff. 1960. *A History of Civilization*. 2 vols. 2nd ed. Englewood Cliffs, NJ: Prentice-Hall.

Broad, William J. "It Swallowed a Civilization." *New York Times* (October 21, 2003), D1–2.

Brown, Donald E. 1991. *Human Universals*. Philadelphia: Temple University Press.

Bryson, Bill. 2003. *A Short History of Nearly Everything*. New York: Broadway Books.

Budiansky, Stephen. 1992. *The Covenant of the Wild: Why Animals Chose Domestication*. New York: William Morrow.

Bulliet, Richard W., et al. 2003. *The Earth and Its People: A Global History*. Brief 2nd ed. Boston: Houghton Mifflin.

Campbell, Joseph, with Bill Moyers. 1988. *The Power of Myth*. New York: Doubleday.

Capra, Fritjóf. 2002. *The Hidden Connections: Integrating the Biological, Cognitive, and Social Dimensions of Life into a Science of Sustainability*. New York: Doubleday.

Capra, Fritjóf and David Steindl-Rast. 1991. *Belonging to the Universe: Explorations on the Frontiers of Science and Spirituality*. New York: HarperCollins.

Carson, I.A. Ritchie. 1981. *Food in Civilization: How History Has Been Affected by Human Tastes*. New York and Toronto: Beaufort Books.

Cavalli-Sforza, Luigi Luca. 2000. *Genes, Peoples, and Languages*. Translated from the Italian by Mark Seielstad. New York: North Point/Farrar, Straus and Giroux.

Christian, David. "The Case for Big History." *Journal of World History* 2, no. 2 (Fall 1991): 223–238. (Reprinted in *The New World History: A Teacher's Companion*. Ross E. Dunn, ed. Boston: Bedford/St. Martin's, 2000, 575–587.)

———. 1997. *Imperial and Soviet Russia: Power, Privilege and the Challenge of Modernity*. New York: St. Martin's Press.

———. 1998. *A History of Russia, Central Asia and Mongolia. Vol. 1, Inner Eurasia from Prehistory to the Mongol Empire*. Oxford: Blackwell.

———. "World History in Context." *Journal of World History* 14, no. 4 (December 2003): 437–488.

———. 2004. *Maps of Time: An Introduction to Big History*. Berkeley: University of California Press.

Cohen, Mark Nathan. 1977. *The Food Crisis in Prehistory: Overpopulation and the Origins of Agriculture*. New Haven: Yale University Press.

Cohen, Shaye J.D. 1988. "Roman Domination: The Jewish Revolt and the Destruction of the Second Temple." In Hershel Shanks, ed. *Ancient Israel: A Short History from Abraham to the Roman Destruction of the Temple*. Englewood Cliffs, NJ: Prentice-Hall, 205–235.

Collins, Roger. 1998. *Charlemagne*. Toronto and Buffalo: University of Toronto Press.

Cook, Michael. 2003. *A Brief History of the Human Race*. New York: W. W. Norton.

Cotterell, Arthur, ed. 1980. *The Penguin Encyclopedia of Ancient Civilizations*. London and New York: Penguin Books.

Crawford, Harriet. 1991. *Sumer and the Sumerians*. Cambridge: Cambridge University Press.

Curtin, Philip D. 1984. *Cross-Cultural Trade in World History*. Cambridge: Cambridge University Press.

D'Altroy, Terence N. 2002. *The Incas*. Malden, MA: Blackwell Publishers.

Davidson, Basil. 1991. *African Civilization Revisited*. Trenton, NJ: Africa World Press.

Davies, Nigel. 1995. *The Incas*. Niwot, CO: University Press of Colorado.

Davis, Mike. 2001. *Late Victorian Holocausts: El Niño Famines and the Making of the Third World*. London and New York: Verso.

Dawkins, Richard. 2004. *The Ancestor's Tale: A Pilgrimage to the Dawn of Evolution*. Boston: Houghton Mifflin.

Demandt, Alexander. 1984. *Der Fall Roms: Die Auflösung des römischen Reiches im Urteil der Nachweld*. Munich: Beck.

Desowitz, Robert S. 1997. *Who Gave Pinta to the Santa Maria? Tracking the Devastating Spread of Lethal Tropical Disease into America*. New York: Harcourt Brace.

De Waal, Frans, ed. 2001. *Tree of Origin: What Primate Behavior Can Tell Us About Human Social Evolution*. Cambridge, MA: Harvard University Press.

De Waal, Frans. 2005. *Our Inner Ape: A Leading Primatologist Explains Why We Are Who We Are*. New York: Riverhead Books.

Diamond, Jared. 1991. *The Rise and Fall of the Third Chimpanzee*. London: Radius.

———. 1999. *Guns, Germs and Steel: The Fates of Human Societies*. New York: Norton.

———. 2005. *Collapse: How Societies Choose to Fail or Succeed*. New York: Viking Penguin.

Dickinson, Terence. 1992. *The Universe and Beyond*. Revised and expanded. Buffalo, NY: Camden.

Dunn, Ross E. 1986. *The Adventures of Ibn Battuta: A Muslim Traveler of the Fourteenth Century*. Berkeley: University of California Press.

Easterbrook, Gregg. 1995. *A Moment on Earth: The Coming Age of Environmental Optimism*. New York: Penguin.

Eaton, Richard M. 1990. *Islamic History as Global History*. Washington, DC: American Historical Association.

Ehrenberg, Margaret. 1989. *Women in Prehistory*. Norman: University of Oklahoma Press.

Eisler, Riane. 1987. *The Chalice and the Blade: Our History, Our Future*. New York: Harper and Row.

Ellis, Peter Bernesford. 1990. *The Celtic Empire: The First Millennium of Celtic History c. 1000 B.C.–51 A.D.* London: Constable.

Erwin, Douglas H. 2006. *Extinction: How Life on Earth Nearly Ended 250 Million Years Ago*. Princeton, NJ: Princeton University Press.

Fagan, Brian M. 1984. *The Aztecs*. New York: W. H. Freeman.

———. 1990. *The Journey from Eden: The Peopling of Our World*. London: Thames and Hudson.

———. 1992. *People of the Earth: An Introduction to World Prehistory*, 7th ed. New York: HarperCollins.

———. 2005. *The Long Summer: How Climate Changed Civilization*. New York: Basic Books.

Fernandez-Armesto, Felipe. 1995. *Millennium: A History of the Last Thousand Years*. New York: Simon and Schuster.

———. 2003. *The Americas: A Hemispheric History*. New York: Modern Library.

Ferris, Timothy. 1997. *The Whole Shebang: A State-of-the-Universe Report.* New York: Simon and Schuster.

Finlay, Robert. "How Not to (Re)Write World History: Gavin Menzies and the Chinese Discovery of America." *Journal of World History* 15, no. 2 (June 2004), 229–242.

Fouts, Roger, with S.T. Mills. 1997. *Next of Kin: My Conversations with Chimpanzees.* New York: Avon.

Frederickson, George M. 2002. *Racism: A Short History,* Princeton, NJ: Princeton University Press.

Gallup, George Jr., and D. Michael Lindsay. 1999. *Surveying the Religious Landscape: Trends in U.S. Beliefs.* Harrisburg, PA: Morehouse Publishing.

Gladwell, Malcolm. 2000. *The Tipping Point: How Little Things Can Make a Big Difference.* Boston: Bay Books; Little, Brown.

Golden, Peter B. 2001. "Nomads and Sedentary Societies in Eurasia." In Michael Adas, ed., *Agricultural and Pastoral Societies in Ancient and Classical History.* Philadelphia: Temple University Press, 71–115.

Gonick, Larry. 1990. *The Cartoon History of the Universe: From the Big Bang to Alexander the Great.* New York: Doubleday.

Goodenough, Ursula. 1998. *The Sacred Depths of Nature.* New York and Oxford: Oxford University Press.

Goudsblom, Johan, Eric Jones, and Stephen Mennell. 1996. *The Course of Human History: Economic Growth, Social Process and Civilization.* Armonk, NY: M.E. Sharpe.

Gould, Stephen Jay. "Down on the Farm: A Review of Donald O. Henry, *From Foraging to Agriculture: The Levant at the End of the Ice Age.*" *New York Review of Books* (January 18, 1990), 26–27.

Gould, Stephen Jay, ed. 1993. *The Book of Life: An Illustrated History of the Evolution of Life on Earth.* New York: W. W. Norton.

Greene, Brian. 1999. *The Elegant Universe.* New York: Vintage.

Hannaford, Ivan. 1996. *Race: The History of an Idea in the West.* Washington, DC: Woodrow Wilson Center Press.

Hansen, Jim. "The Threat to the Planet." *New York Review of Books,* July 13, 2006, 12–14, 16.

Harris, Sam. 2004. *The End of Faith: Religion, Terror and the Future of Reason.* New York: W. W. Norton.

Heiser, Charles B. Jr. 1981. *Seed to Civilization: The Story of Food,* 2nd ed. San Francisco: W. H. Freeman.

Hemming, John. 1970. *The Conquest of the Incas.* New York: Macmillan.

Henry, Donald O. 1989. *From Foraging to Agriculture: The Levant at the End of the Ice Age.* Philadelphia: University of Pennsylvania Press.

Hick, John. 1989. *An Interpretation of Religion: Human Responses to the Transcendent.* New Haven: Yale University Press.

Hillel, Daniel J. 1991. *Out of the Earth: Civilization and the Life of the Soil.* New York: Free Press, Macmillan.

Howard, W. J. 1991. *Life's Beginnings.* Coos Bay, OR: Coast Publishing.

Hughes, J. Donald. 1975. *Ecology in Ancient Civilizations.* Albuquerque: University of New Mexico.

Hughes, J. Donald, ed. 2000. *The Face of the Earth: Environment and World History.* Armonk, NY: M.E. Sharpe.

Hughes, Sarah Shaver, and Brady Hughes. 2001. "Women in Ancient Civilizations." In

Michael Adas, ed., *Agricultural and Pastoral Societies in Ancient and Classical History*. Philadelphia: Temple University Press, 116–150.

Iliffe, John. 1995. *Africans: The History of a Continent*. New York: Cambridge University Press.

James, Edward. 1988. *The Franks*. London and New York: Blackwell.

Jaspers, Karl. 1953. *The Origin and Goal of History*. Translated by Michael Bullock. New Haven: Yale University Press.

Jay, Peter. 2000. *The Wealth of Man*. New York: Public Affairs.

Jean, Georges. 1992. *Writing: the Story of Alphabets and Scripts*. Translated from the French by Jenny Oates. New York: Harry N. Abrams.

Johns, Catherine. 1982. *Sex or Symbol? Erotic Images of Greece and Rome*. London: British Museum Press.

Jones, Gwyn. 1984. *A History of the Vikings*, rev. ed. Oxford and New York: Oxford University Press.

Keiger, Dale. "Clay, Paper, Code." *Johns Hopkins Magazine*. (September 2003), 34–41.

Kennedy, Paul. 1991. *Preparing for the Twenty-first Century*. New York: Ballantine.

Kicza, John E. 2001. "The People and Civilizations of the Americas Before Contact." In Michael Adas, ed. *Agricultural and Pastoral Societies in Ancient and Classical History*. Philadelphia: Temple University Press, 183–223.

Kilgour, Frederick. 1998. *The Evolution of the Book*. New York: Oxford University Press.

Kivel, Paul. 2004. *You Call This a Democracy? Who Benefits, Who Pays, and Who Really Decides*. New York: Apex Press.

Kirshner, Robert T. 2003. *The Extravagant Universe: Exploding Stars, Dark Energy and the Accelerating Cosmos*. Princeton, NJ: Princeton University Press.

Knight, Franklin W. "Black Athena." *Journal of World History* 4, no. 2 (Fall 1993), 325–327.

Kolbert, Elizabeth. "Why Work?" *New Yorker* (November 29, 2004), 154–160.

Kurlansky, Mark. 2002. *Salt: A World History*. New York: Walker.

Kurtz, Lester. 1995. *Gods in the Global Village: The World's Religions in Sociological Perspective*. Thousand Oaks, CA: Pine Forge Press.

Leakey, Richard, and Roger Lewin. 1995. *The Sixth Extinction: Patterns of Life and the Future of Mankind*. New York: Doubleday.

Levathes, Louise. 1994. *When China Ruled the Seas: The Treasure Fleet of the Dragon throne, 1405–1433*. New York: Simon and Schuster.

Lewin, Roger. 1988. *In the Age of Mankind*. Washington, DC: Smithsonian Books.

———. 1999. *Human Evolution: An Illustrated Introduction*, 4th ed. Malden, MA: Blackwell Science.

Liu, Xinru. 2001. "The Silk Road: Overland Trade and Cultural Interactions in Eurasia." In Michael Adas, ed., *Agricultural and Pastoral Societies in Ancient and Classical History*. Philadelphia: Temple University Press, 151–179.

Lomborg, Bjorn. 2001. *The Skeptical Environmentalist: Measuring the Real State of the World*. Cambridge: Cambridge University Press.

Love, Spencie. 1996. *One Blood: The Death and Resurrection of Charles R. Drew*. Chapel Hill, NC: University of North Carolina Press.

Lovejoy, Paul E. 2000. *Transformations in Slavery: A History of Slavery in Africa*, 2nd ed. Cambridge: Cambridge University Press.

Lovelock, James. 1979. *Gaia: A New Look at Life on Earth*. Reprint, Oxford: Oxford University Press, 1987.

——. 1988. *The Ages of Gaia: A Biography of Our Living Earth*. New York: Bantam Books.

MacMullen, Ramsey. 1984. *Christianizing the Roman Empire (A.D. 100–400)*. New Haven and London: Yale University Press.

McNeill, J.R. 2000. *Something New Under the Sun: An Environmental History of the Twentieth-Century World*. New York: W. W. Norton.

McNeill, J. R. and William H. McNeill. 2003. *The Human Web: A Bird's-Eye View of World History*. New York: W. W. Norton.

McNeill, William H. 1976. *Plagues and People*. Garden City, NJ: Anchor Press/ Doubleday.

——. 1992. *The Global Condition: Conquerors, Catastrophes and Community*. Princeton, NJ: Princeton University Press.

——. "New World Symphony." *New York Review of Books* (December 1, 2005), 43–45.

——. "Secrets of the Cave Paintings." *New York Review of Books* (October 19, 2006), 20–23.

Mann, Charles C. 2005. *1491: New Revelations of the Americas Before Columbus*. New York: Alfred A. Knopf.

Manning, Patrick. 1990. *Slavery and African Life: Occidental, Oriental and African Trade*. Cambridge and New York: Cambridge University Press.

Margulis, Lynn, and Dorion Sagan. 1986. *Microcosmos: Four Billion Years of Evolution from Our Microbial Ancestors*. Berkeley: University of California Press; rev. ed. 1997.

Marks, Jonathan. 2002. *What It Means to Be 98% Chimpanzee: Apes, People, and Their Genes*. Berkeley: University of California Press.

Markus, Robert Austin. 1974. *Christianity in the Roman World*. London: Thames and Hudson.

Meadows, Donella, et al. *The Limits to Growth: A Report for the Club of Rome's Project on the Predicament of Mankind*. New York: Universe Books.

Meadows, Donella, Jorgen Randers, and Dennis Meadows. 2004. *Limits to Growth: The 30-Year Update*. White River Junction, VT: Chelsea Green Publishing.

Mears, John A. 2001. "Agricultural Origins in Global Perspective." In Michael Adas, ed., *Agricultural and Pastoral Societies in Ancient and Classical History*. Philadelphia: Temple University Press, 36–70.

Mellaart, James. 1967. *Çatal Hüyük: A Neolithic Town in Anatolia*. New York: McGraw-Hill.

Meltzer, Milton. 1990. *Columbus and the World Around Him*. New York: Franklin Watts.

Menzies, Gavin. 2002. *1421: The Year China Discovered America*. New York: William Morrow.

Miles, Rosalind. 1990. *The Women's History of the World*. New York: Harper and Row, 1990.

Miller, Walter. 1959. *A Canticle for Leibowitz*. New York: Bantam, reprint 1997.

Mitchell, Stephen. 2004. *Gilgamesh: A New English Version*. New York: Free Press.

Mithen, Stephen. 1996. *The Prehistory of the Mind: The Cognitive Origins of Art, Religion and Science*. London: Thames and Hudson.

Moore, Stephen, and Julian L. Simon. 2000. *It's Getting Better All the Time: 100 Greatest Trends of the Last 100 Years*. Washington, DC: Cato Institute.

Morgan, David. 1986. *The Mongols*. Oxford: Basil Blackwell.

Morris, Craig, and Adriana von Hagen. 1993. *The Inka Empire and its Andean Origins.* New York: Abbeville Press and the American Museum of Natural History.

Officer, Charles, and Jake Page. 1993. *Tales of the Earth: Paroxysms and Perturbations of the Blue Planet.* New York: Oxford University Press.

Pacey, Arnold. 1990. *Technology in Civilization: A Thousand-Year History.* Cambridge, MA: MIT Press.

Patterson, Thomas C. 1991. *The Inca Empire: The Formation and Disintegration of a Pre-Capitalist State.* New York and Oxford: Berg, St. Martin's Press [distributor].

Pearce, Fred. 2006. *When Rivers Run Dry.* Boston: Beacon.

Pinker, Stephen. 1994. *The Language Instinct: How the Mind Creates Language.* New York: William Morrow; 2000, HarperCollins Perennial.

———. 1997. *How the Mind Works.* New York: W.W. Norton.

Pollan, Michael. 2001. *Botany of Desire: A Plant's Eye View of the World.* New York: Random House.

Pomeranz, Kenneth. 2000. *The Great Divergence: China, Europe, and the Making of the Modern World Economy.* Princeton, NJ: Princeton University Press.

Ponting, Clive. 1991. *A Green History of the World: The Environment and the Collapse of Great Civilizations.* New York: Penguin.

Potts, Timothy. "Buried Between the Rivers." *New York Review of Books* (September 25, 2003), 18–23.

Prantzos, Nikos. 2000. *Our Cosmic Future: Humanity's Fate in the Universe.* Cambridge: Cambridge University Press.

Ratchnevsky, Paul. 1991. *Genghis Khan: His Life and Legacy.* Trans. T. N. Haining. Oxford: Blackwell.

Rees, Martin. 2003. *Our Final Hour: A Scientist's Warning: How Terror, Error, and Environmental Disaster Threatens Humankind's Future in this Century—On Earth and Beyond.* New York: Basic Books.

Relethford, John H. 1994. *The Human Species: An Introduction to Biological Anthropology,* 2nd ed. Mountain View, CA: Mayfield Publishing.

Reston, James Jr. 2005. *Dogs of God: Columbus, the Inquisition, and the Defeat of the Moors.* New York: Doubleday.

Richards, John F. 2003. *The Unending Frontier: An Environmental History of the Early Modern World.* Berkeley: University of California Press.

Ridley, Mark. 1996. *The Origins of Virtue: Human Instincts and the Evolution of Cooperation.* New York: Viking Penguin.

Robinson, Kim S. 1991. *Red Mars.* New York: Bantam.

———. 1994. *Green Mars.* New York: Bantam.

———. 1996. *Blue Mars.* New York: Bantam.

Rollins, Allen M. 1983. *The Fall of Rome: A Reference Guide.* Jefferson, NC: McFarland and Company.

Ronan, Colin A. 1978. *The Shorter Science and Civilization in China: An Abridgement of Joseph Needham's Original Text,* vol. 1. Cambridge, England: Cambridge University Press.

Rue, Loyal. 2000. *Everybody's Story: Wising Up to the Epic of Evolution.* New York: State University of New York Press.

Ruhlen, Merritt. 1994. *The Origin of Language: Tracing the Evolution of the Mother Tongue.* New York: John Wiley.

Ryan, William, and Walter Pitman. 1999. *Noah's Flood: The New Scientific Discoveries About the Event That Changed History*. New York: Simon and Schuster.

Sahlins, Marshall. 1972. *Stone Age Economics*. Chicago and New York: Aldine-Atherton, Inc.

Sales, Kirkpatrick. 1990. *The Conquest of Paradise: Christopher Columbus and the Columbian Legacy*. New York: Alfred A. Knopf.

Sands, Roger. 2005. *Forestry in a Global Context*. Cambridge, MA: CABI Publishing.

Sapolsky, Robert. 2001. *A Primate's Memoir: A Neuroscientist's Unconventional Life Among the Baboons*. New York: Simon and Schuster.

Schulman, Erik. 1999. *A Briefer History of Time: From the Big Bang to the Big Mac*. New York: W. H. Freeman.

Shanklin, Eugenia. 1994. *Anthropology and Race*. Belmont, CA: Wadsworth.

Shanks, Hershel, ed. 1988. *Ancient Israel: A Short History from Abraham to the Roman Destruction of the Temple*. Englewood Cliffs, NJ: Prentice-Hall.

Skidelsky, Robert. "The Mystery of Growth." *New York Review of Books* (March 13, 2003), 28–31.

Smil, Vaclav. 1994. *Energy in World History*. Boulder, CO: Westview Press.

Smith, Alan K. 1991. *Creating a World Economy: Merchant Capital, Colonialism and World Trade, 1400–1825*. Boulder, CO: Westview Press.

Smith, Huston. 1991. *The World's Religions*, rev. ed. New York: HarperCollins.

Smith, Huston, and Philip Novak. 2003. *Buddhism: A Concise Introduction*. New York: HarperCollins.

Smolin, Lee. 1998. *The Life of the Cosmos*. London: Phoenix.

Spier, Fred. 1996. *The Structure of Big History: From the Big Bang Until Today*. Amsterdam: Amsterdam University Press.

Standford, Craig. 2001. *Significant Others: The Ape-Human Continuum and the Quest for Human Nature*. New York: Basic Books.

Stavrianos, L.S. 1989. *Lifelines from Our Past: A New World History*. New York: Pantheon Books.

Stearns, Peter N. 1987. *World History: Patterns of Change and Continuity*. New York: Harper and Row.

———. 1993. *The Industrial Revolution in World History*. Boulder, CO: Westview Press.

Sterngold, James. "Experts Fear Nuke Genie's Out of Bottle." *San Francisco Chronicle* (November 22, 2004), A1, 8.

Stuart, George. 1949. *Earth Abides*. New York: Random House.

Summer Institute of Linguistics. 1990. *The Alphabet Makers*. Huntington Beach, CA: Summer Institute of Linguistics.

Swimme, Brian, and Thomas Berry. 1992. *The Universe Story: From the Primordial Flaring Forth to the Ecozoic Era: A Celebration of the Unfolding of the Cosmos*. San Francisco: Harper San Francisco.

Sykes, Bryan. 2001. *The Seven Daughters of Eve: The Science That Reveals Our Genetic Ancestry*. New York: W. W. Norton.

Tainter, Joseph A. 1989. *The Collapse of Complex Societies*. Cambridge: Cambridge University Press.

Temple, Robert. 1986. *The Genius of China: Three Thousand Years of Science, Discovery and Invention*. New York: Simon and Schuster.

Thomas, Hugh. 2004. *Rivers of Gold: The Rise of the Spanish Empire, from Columbus to Magellan*. New York: Random House.

Thomson, Hugh. 2001. *The White Rock: An Exploration of the Inca Heartland.* Woodstock and New York: Overlook Press.

Tilly, Louise A. 1993. "Industrialization and Gender Inequality." In Michael Adas, ed. *Essays on Global and Comparative History.* Washington, DC: American Historical Association.

Todd, Ian A. 1976. *Çatal Hüyük in Perspective.* Menlo Park, CA: Cummings Publishing.

Tudge, Colin. 1996. *The Time Before History: Five Million Years of Human Impact.* New York: Scribner.

———. 1999. *Neanderthals, Bandits, and Farmers: How Agriculture Really Began.* New Haven and London: Yale University Press.

Turville-Petre, E.O.G. 1975. *Myth and Religion of the North: The Religion of Ancient Scandinavia.* Westport, CT: Greenwood Press.

Tyerman, Christopher. 2006. *God's War: A New History of the Crusades.* Cambridge, MA: Harvard University Press.

Van Doren, Charles. 1991. *A History of Knowledge: Past Present and Future.* New York: Ballantine.

Van Sertima, Ivan. 1976. *They Came Before Columbus.* New York: Random House.

———. 1998. *Early America Revisited.* Piscataway, NJ: Transaction Publishers.

Van Sertima, Ivan, ed. 1992. *African Presence in Early America.* New Brunswick, NJ: Transaction Books.

Wade, Nicolas. "In Click Languages an Echo of the Tongues of the Ancients." *New York Times* (March 18, 2003), science section.

Ward-Perkins, Bryan. 2005. *The Fall of Rome and the End of Civilization.* Oxford: Oxford University Press.

Weatherford, Jack. 1988. *Indian Givers: How the Indians of the Americas Transformed the World.* New York: Fawcett Columbine.

———. 1991. *Native Roots: How the Indians Enriched America.* New York: Crown Publishers.

———. 2004. *Genghis Khan and the Making of the Modern World.* New York: Crown Publishers.

White, Randall. 1986. *Dark Caves, Bright Visions.* New York: American Museum of Natural History.

Wilford, John Noble. "9000-Year-Old Cloth Found." *San Francisco Chronicle* (August 13, 1993).

Willets, R.F. 1980. "The Minoans." In Arthur Cotterell, ed., *The Penguin Encyclopedia of Ancient Civilizations.* London and New York: Penguin, 204–210.

Williams, Michael. 2003. *Deforesting the Earth: From Prehistory to Global Crisis.* Chicago: University of Chicago Press.

Wilson, David M. 1989. *The Vikings and Their Origins: Scandinavia in the First Millennium,* rev. ed. London: Thames and Hudson.

Wilson, Edward O. 2006. *The Creation: An Appeal to Save Life on Earth.* New York: W. W. Norton.

Wilson, Ian. 2001. *Past Lives: Unlocking the Secrets of Our Ancestors.* London: Cassell.

Wolkstein, Diane, and Samuel Noah Kramer. 1983. *Inanna: Queen of Heaven and Earth: Her Stories and Hymns from Sumer.* New York: Harper and Row.

Wrangham, Richard W. 2001. "Out of the *Pan*, into the Fire: How Our Ancestors' Evolution Depended on What They Ate." In Frans B. M. De Waal, ed., *Tree of Origin:*

What Primate Behavior Can Tell Us About Human Social Evolution. Cambridge, MA: Harvard University Press, 121–143.

Yoffee, Norman, and George L. Cowgill, eds. 1988. *The Collapse of Ancient States and Civilizations.* Tucson, AZ: University of Arizona Press.

Zinn, Howard. 1980. *A People's History of the United States.* New York: Harper and Row.

Index

CELEBRATING 20 YEARS OF INDEPENDENT PUBLISHING

Thank you for reading this book published by The New Press. The New Press is a nonprofit, public interest publisher celebrating its twentieth anniversary in 2012. New Press books and authors play a crucial role in sparking conversations about the key political and social issues of our day.

We hope you enjoyed this book and that you will stay in touch with The New Press. Here are a few ways to stay up to date with our books, events, and the issues we cover:

- Sign up at www.thenewpress.com/subscribe to receive updates on New Press authors and issues and to be notified about local events
- Like us on Facebook: www.facebook.com/newpressbooks
- Follow us on Twitter: www.twitter.com/thenewpress

Please consider buying New Press books for yourself; for friends and family; or to donate to schools, libraries, community centers, prison libraries, and other organizations involved with the issues our authors write about.

The New Press is a 501(c)(3) nonprofit organization. You can also support our work with a tax-deductible gift by visiting www.thenewpress.com/donate.